The Mountains of New Mexico

The author at the mouth of Shipman Canyon in the San Mateo Mountains.

The Mountains of New Mexico

Robert Julyan

Photographs by Carl Smith

University of New Mexico Press • Albuquerque

©2006 by the University of New Mexico Press
All rights reserved. Published 2006
12 11 10 09 08 07 06   1 2 3 4 5 6 7

Library of Congress Cataloging-in-Publication Data

Julyan, Robert Hixson.
The mountains of New Mexico / Robert Julyan ; photographs by Carl Smith.
p. cm.
Includes bibliographical references and index.
ISBN-13: 978-0-8263-3515-9 (cloth : alk. paper)
ISBN-10: 0-8263-3515-2 (cloth : alk. paper)
ISBN-13: 978-0-8263-3516-6 (pbk. : alk. paper)
ISBN-10: 0-8263-3516-0 (pbk. : alk. paper)
1. Mountains—New Mexico.
2. New Mexico—Description and travel.
3. Natural history—New Mexico. I. Title.
F802.A16J85 2006
917.89—dc22

2005037317

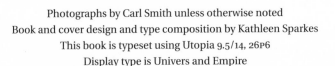

Photographs by Carl Smith unless otherwise noted
Book and cover design and type composition by Kathleen Sparkes
This book is typeset using Utopia 9.5/14, 26P6
Display type is Univers and Empire

To Mike and to the mountains
we roamed together as boys.

Contents

Maps

Preface

*"My final prayer was one of thanksgiving for a world
filled with the sublimity of the high places, for the sheer
beauty of the mountains, and for the surpassing miracle
that we should be so formed as to respond
with ecstasy to such beauty...."*

—Mountaineer Willi Unsoeld, upon the death of
his daughter, Nanda Devi, while climbing her
namesake mountain in the Himalayas.

grew up in mountains and have loved them all my life. That's why I wrote
this book. Well, that and the fact that I needed another excuse to continue
rambling around my adopted state after I'd completed my earlier books
about the state's place names, wilderness areas, and hiking trails. There's just
something habit-forming about getting in a car, or setting off on a trail, to
see what reality lies beneath the names and spaces on maps or text in books.

And that's what I invite you to do with this book. I don't for a moment
believe I did more than provide a brief overview of New Mexico's mountains, but I do hope I'll whet your curiosity as to what the mountains
included here are really like, and that you'll be moved to get out and explore
as I have. I promise that if you do you'll not be disappointed.

But I must caution you to put your preconceptions aside. If there's anything that unites New Mexico's mountains it's that they never are what a
mere superficial impression would suggest. I can't tell you how many times
I've gone to an unprepossessing bump on the landscape and discovered
there wonders, delights, and even mysteries completely unexpected. In fact,

I half-facetiously say that New Mexico is a place where you can throw a dart at a map, and know that wherever it sticks will be something interesting.

What's more, your dart has a great chance of hitting a mountain. When I told friends in Colorado I was writing a book about New Mexico's mountains, I sometimes received a puzzled, skeptical look that seemed to say, "It must be a slim book." I wasn't totally surprised. Sooner or later everyone who lives in New Mexico must confront non-residents' misperception that we're denizens of the desert, cousins of Gila monsters and javelinas, scurrying among sand dunes and giant cacti. But the reality is that New Mexico is very much a mountain state. It has more than one hundred named mountain groups, surpassing Colorado's total, and while Colorado's mountains far exceed New Mexico's mountains in elevation and what people expect of mountain scenery, New Mexico's highest point, 13,161-foot Wheeler Peak, exceeds the highest points in Idaho (Borah Peak, 12,662), Montana (Granite Peak, 12,799), Nevada (Boundary Peak, 13,140), Oregon (Mount Hood, 11,239), and Arizona (Humphreys Peak, 12,633), and is only slightly lower than those in Utah (Kings Peak, 13,528) and Wyoming (Gannett, 13,804). Thus, when ranked by highest peak, New Mexico is the fifth highest state. (It's also the nation's fifth largest state.)

Moreover, while pockets of New Mexico do conform to desert stereotypes, true desert landscapes (sand dunes, giant cacti, etc.) are uncommon here. To be sure, of the four deserts found in the United States—Mohave, Sonoran, Great Basin, and Chihuahuan—two exist in New Mexico: the Great Basin in the northwestern part of the state and the Chihuahuan throughout southern New Mexico. But even the desert regions here have mountains. Just to the east of the Tularosa Basin, where the shifting dunes of White Sands create as ideal a desert landscape as one would imagine, rises Sierra Blanca, 11,973 feet high, snow-capped for most of the year and the most southerly mountain in North America glaciated during the last Ice Age. Only New Mexico's far southeast and east-central regions lack significant mountain features.

Also, New Mexico's mountains span an enormous range of geography and ecology. All of New Mexico's five physiographic provinces include mountain features: glacier-carved alpine summits in the Sangre de Cristos, composite volcanoes such as Mount Taylor and Sierra Grande, cinder cones such as Capulin Mountain, fossil limestone reefs in the Guadalupes, laccolith intrusions in the Capitan and Zuni Mountains, erosional formations such as Tucumcari Mountain, and tilted fault-blocks in the Sandias and Caballos.

Concomitant with this is a comparable ecological diversity; it is impossible to speak of a single mountain fauna here, as in the northern Rocky Mountains, for mountain animals here range from elk in the high alpine meadows of the Latir Peaks to desert bighorn sheep in desert ranges such as the Big Hatchets, from marmots in the tundra of Wheeler Peak to coatimundis prowling in Peloncillo arroyos. Within New Mexico have been sighted arctic lynx and semi-tropical jaguars.

The human history of New Mexico's mountains similarly defies easy characterization. Spanning at least 11,500 years, humans here have been hunter-gatherers, agriculturalists, prospectors and miners, loggers, ranchers, hunters and anglers, scientists, hikers, climbers, mountain-bikers, artists, and even photographers and writers. That, too, is among the joys of exploring New Mexico's mountains—being jolted out of the present by reminders that others have been here before. I recall making a very difficult scramble to the top of Cox Peak, in the arid void southwest of Las Cruces and finding there a crude stone circle. Who made it? Why? How long ago? And what did they behold? Tantalizing questions everywhere.

But it's a truism of exploration that the more you know, the more you discover. So again, I hope this book will expand your awareness as you go about making your own discoveries.

Acknowledgments

A book such as this could not exist without the help and cooperation of many people, and to all of them I am deeply grateful. I'm especially grateful to Dave Love and the other people at the NM Bureau of Geology and Mineral Resources—Jane Love, Robert Eveleth, David McCraw, L. Greer Price—and to their director, Peter Scholle, for allowing me access to the resources of his staff. Dave Love's generous sharing of his time and knowledge was truly invaluable. Similarly, I'm grateful to Mary Wyant of the University of New Mexico Map and Geographic Information Center for making available the skills and time of Knutt Peterson, who undertook the challenge of creating the maps in this book. GIS specialist Kurt Menke also helped with the maps. Tom Ferguson, longtime NM Mountain Club member, shared his knowledge and keen eyes in reviewing the manuscript, as did Carl Smith. This was but one of the many ways in which Carl contributed to this project beyond taking the photographs; I'll always treasure the memories of our adventures together exploring the far-flung mountains of New Mexico, and when I encountered a dilemma in my writing he always was the first person to whom I turned. And finally, I appreciate deeply the support and experience of the staff of the University of New Mexico Press, especially David Holtby, Kathy Sparkes, Maya Allen-Gallegos, Sonia Dickey, Sarah Ritthaler, and Lisa Pacheco; no author ever had more patient and understanding publishers.

To all these people and more, including my wife, Mary, who accepted me taking off to places she'd never heard of, I am deeply grateful.

New Mexico's Roads and Towns and Mountains

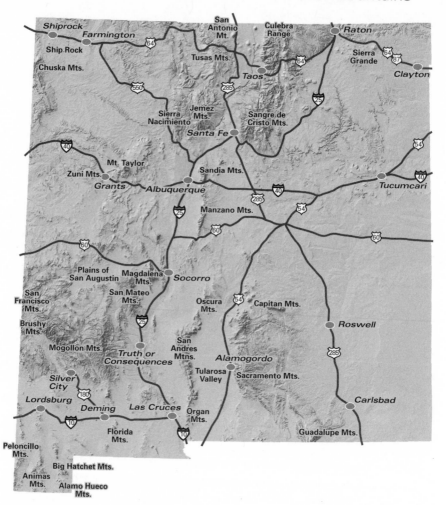

Introduction

How to Use This Book

The mountains in this book are organized by the state's physiographic provinces, whose characteristics are summarized in this introduction and more fully at the beginning of each section. Each feature is introduced by an information block.

Location: Here the feature's location is described relative to widely-recognized landmarks—towns, highways, water bodies, and others—where possible.

Physiographic province: This identifies the geologic and topographic region of the state where the feature is located.

Elevation range: This gives a general idea of the range's relief, which is the difference between the elevation at a mountain's base and its highest elevation. For individual mountains I provide relief rather than elevation range. The upper end of the elevation range doesn't necessarily correlate with a mountain group's highest point, which can be much higher than most other summits in the range, and neither figure was derived by any method more precise than looking at a topographic map. Knowing a range's elevation is important because elevation strongly influences weather and vegetation.

High point: This is simply the highest elevation in the mountain group, regardless of whether it's atop a mountain, a mesa, or an upland; regardless of whether it's named or unnamed. The elevations were obtained from U.S. Geological Survey (USGS) topographic maps and from the USGS Geographic Names Information System (GNIS) database. These sources are regarded as official for all federal maps. Sometimes, however, a lag exists between surveying and mapping. Where the high point of a range is in another state or country, such as the Guadalupe Mountains and the Sierra Rica, I have attempted to include the entire range's high point and also the high point in New Mexico.

Other major peaks: These invariably are named. Usually, they owe their prominence to elevation and size, but a summit can also attain importance as a landmark or being near a population center.

Dimensions: Most commonly the boundaries of individual mountains and mountain ranges are obvious, but not always. For example, the boundaries of the many sub-ranges of the Sangre de Cristo Mountains are very indistinct. What's more, no authority usually exists to which one can turn to verify boundaries. Occasionally, boundaries are defined by the USGS, as when the issue arose as to the extent of the Tusas Mountains, west of US 285. Geology also defines boundaries, as in the Jemez Mountains and the contiguous Sierra Nacimiento. But more likely such adjudication has not occurred, and one is left simply with a vague, intuitive sense of where one range stops and another begins.

In this book I've not presumed to apply any strict criteria to resolve these ambiguities but rather have relied upon my best judgment to estimate mountain dimensions.

Ecosystems: This is just a brief listing of the dominant and common plants in the mountains (see *Ecosystems and Life Zones*).

Counties: When listing the counties in which the range occurs, the first county is the one encompassing the range's highest point.

Administration: This refers to landownership and management. It's common for a range to be overlain by a patchwork of ownership, including the U.S. Forest Service (USFS), U.S. Bureau of Land Management (BLM), National Park Service (NPS), state agencies, and private owners.

Wilderness: This refers only to formally declared or proposed wildernesses (Wilderness Study Areas) and is *not* a reflection of how wild a range might actually be. Some remote mountains might have no formal wildernesses but be very wild and natural indeed.

Getting there: Here I present what I believe to be a clear and efficient route for approaching the mountains and exploring them; not necessarily the most scenic or interesting route, simply a practical one. Regarding road conditions, I've tended to err on the side of caution. Some ranges, such as the Diablo Range, can only be approached by foot. Other ranges, including some large and significant ones, have no public access at all. These directions should not replace a good, current map—conditions are changing constantly. When venturing far from a highway it's always prudent to check with the local land management agencies or local people about current conditions. Sometimes road access through private property is legal as long as you stay on the road; at other times you may encounter a locked gate. Always respect posted signs and private property (see *Land ownership*).

The narrative

Following the information block is a narrative about the mountains. In it I explain the mountains' geology, their natural history, and their human history. In my checklist of things to review for each range were things such as plants and animals, prehistoric peoples and modern Native populations, Hispanic and Anglo history, mining, and any other significant activities. But foremost and always I attempted to identify what about a range makes it individual, imparts a unique character. Thus I felt at liberty to include almost anything that struck my fancy—from interesting rocks to rare plant and animal species to historical anecdotes—and certainly the place names.

This is not a hiking guide. But where appropriate I did try to indicate what hiking exists in the mountains and point toward major trailheads. Also, I tried for each range to describe how one would reach the highest point. Any trail descriptions, however, should not supplant more detailed guides and maps.

Land ownership

In this book I have tried to indicate generally the ownership status of a range or peak, but landownership in New Mexico often is shifting and complex. For example, the Wilderness Study Area that includes the Sierra Ladrones had been a mixture of both state and BLM land, but due to a land exchange all the land within the WSA is now BLM. Such exchanges, as well as other transfers of ownership, are occurring all the time.

Another complication is the "checkerboard" pattern that exists in many parts of the state, wherein the landownership status of alternating square-mile sections is different. Sometimes state land blocks alternate with private land blocks; sometimes the mix is private and BLM, or BLM and state, or BLM and Indian reservation. It can get very complicated, and the configuration constantly is changing. Unfortunately, this pattern means that often you must obtain permission from private landowners to gain access to public land. Because of this, outdoor recreationists in New Mexico often are dependent upon the good will of private landowners and therefore must respect their rights, seek permission when necessary, and obey all signs and notices.

If you are uncertain about the ownership status of land you wish to enter, consult the most current land status map; for example, the *New Mexico Road and Recreation Atlas,* published by Benchmark Maps and available at most bookstores, indicates land status as of publication date.

Names

My source and ultimate reference for names and elevations has been the U.S. Geological Survey (USGS), the federal agency mandated to ensure the accuracy and consistency of both. The USGS is home to the U.S. Board on Geographic Names (BGN) and the Geographic Names Information System (GNIS) database, a larger, multi-agency effort by the federal government to ensure accuracy and consistency in geographic names. By federal law, only the names in GNIS may be used in federal maps and products, including those of the USGS, Bureau of Land Management, U.S. Forest Service, Census Bureau, Government Printing Office, and others. I have been involved with the BGN and GNIS for approximately twenty years, and they were the authorities for official forms of names for my book, *The Place Names of New Mexico.* But as chair of the NM Geographic Names Committee, the official state liaison with the BGN, I'm aware that achieving accuracy and consistency is an ongoing and often complex process. For both the New Mexico names committee and the BGN, the main criterion upon which decisions are made is local usage and preference. In this book, I've used the forms of names listed in GNIS, except were local usage is widely different or where official forms reflect linguistic mistakes, as in the Cerrillos Hills.

Rocky Mountain bighorn sheep, Truchas Peaks

About wildlife

In describing each range's ecology, as well as in providing information for hikers, I have discussed the local wildlife. But much of the information is obvious and repetitive; black bears, coyotes, mule deer, bobcats, even mountain lions, are found throughout the state; hardly a range is without them; they make up the generalized New Mexico fauna. Furthermore, in listing species such as black bears and bobcats again and again, I've had an uneasiness about subtly, unintentionally raising unrealistic expectations of actually seeing these animals. I've lived and hiked in bobcat country most of my life, yet in all that time I've seen but five bobcats—and one as I was driving! As Carl Smith remarked, "My pet peeve: one never sees all the wildlife that guide book writers mention. One suspects it's a list of everything that has been seen in the range for the last 100 years, and has no relevance to what one might see."

Therefore I don't list all the widespread, expected species for each range but focus instead on any fauna of special interest, including:

- Endangered or threatened species in the range, such as the jaguar in the Peloncillo Mountains,

- Species unique to the range, such as the ibex in the Floridas,
- Exceptional abundance of a species, or exceptional viewing opportunities, such as elk in the Valles Caldera in the Jemez Mountains,
- Successful reintroductions, such as desert bighorn in the Magdalena Mountains.

Measurements

Measurements are given in feet; metric conversions are:
one foot = 0.3048 meters
one meter = 3.28 feet.

Mountain Terminology

When I was a child growing up in the Front Range of north-central Colorado, I never imagined that someone might ask, "What is a mountain?" Wasn't it obvious? Mountains are like oceans, fundamental features of the Earth's surface—how could anyone mistake one for something else? Just as you knew you were in the ocean by the limitless horizon and taste of salt water, you knew you were in the mountains by their human-dwarfing size and steep, rocky slopes. But like so many other childhood perceptions, that one proved too simple. The Earth's surface is far more complex than I had suspected—and so is language. Therefore, the question "What is a mountain?" is legitimate.

Especially so in a state such as New Mexico, where nothing, it seems, is simple, including landforms. The Sangre de Cristo Mountains patently are mountains—but what about the Cedar Mountain Range southwest of Deming? The mountains there have gentle, rounded contours, and despite the name, trees of any species are uncommon. Aren't these hills, not mountains, much less a "range," the names notwithstanding?

And while Sierra Blanca, Mount Taylor, Baldy Mountain, and Cookes Peak clearly conform to the stereotypical image of a mountain, what about Mount Dora, "A" Mountain near Las Cruces, and Rabbit Ear Mountain near Clayton? When someone mentions the term *mountain,* these examples do not appear before the mind's eye.

Language is not much help in making decisions as to what is a mountain and what is not. In the Sangre de Cristo Mountains northeast of Taos

Late spring in the Sangre de Cristo Mountains, Gold Hill from the south

there is a 12,716-foot peak, the state's fifteenth highest mountain—and it is called Gold *Hill*. Conversely, west of Jal is a little blip on the landscape, merely 3,232 feet in elevation, less than half a mile in circumference, with relief of just 50 feet—and it's called Custer *Mountain*. An even greater mismatch between name and landform is Railroad Mountain east of Roswell, an igneous dike running more than 15 miles, less than 0.1 mile wide, and seldom more than twenty-five feet high. But the contrast is perhaps best seen in the San Andres Mountains, where the 18-mile-long, eight thousand-foot Chalk Hills are but 3.5 miles from the 2.5-mile-long, sixty-nine hundred-foot Hardscrabble Mountains.

Spanish generic terms are no more precise than the English ones. *Cerro* is the Spanish term for "hill," but in New Mexico the term denotes everything from El Cerro, a volcanic mound near Tomé reaching 5,223 feet in elevation, with 375 feet of relief, to Cerro de la Olla, an 11,932-foot summit in the Sangre de Cristo Mountains, with more than 3,000 feet of relief.

Roderick Peattie, in his classic *Mountain Geography, a Critique and Field Study,* (New York: Greenwood Press, 1936), recognized the difficulty of defining a mountain:

> A mountain, strictly speaking, is a conspicuous elevation of small summit area. A plateau is a similar elevation of large summit area with at least one sheer side. An essential and yet indefinite element in the definition of a mountain is the conspicuity. Conspicuity, like height, is a relative matter, and depends upon the personal evaluation or the standard by which it is measured. . . . Mountains should be impressive; they should enter into the imagination of the people who live within their shadows. Unfortunately, it is next to impossible to include such intangibles in a definition. Mountains have bulk; mountains also have individuality.

Elevation does help. In North America, at least, any feature 13,000 feet or higher is a mountain, but does Organ Peak's 8,870-foot elevation make it less a mountain than, say, 10,000-foot Grass Mountain? Mountaineers harbor no doubts as to Organ Peak's pedigree.

Relief also is important. At 13,161 feet, Wheeler Peak is New Mexico's highest summit, but it has less vertical relief than 11,973-foot Sierra Blanca, which towers 7,000 feet above the Tularosa Basin. Hamilton Mesa in the Pecos Wilderness has little relief, but is 10,000 feet in elevation and covers a large area, whereas Ship Rock is only 7,178 feet and covers a very limited area but has probably the starkest relief of any feature in the state.

Because no criteria have been established to determine what is a mountain and what is not, at least to my knowledge, I have taken the liberty to

Ship Rock, southern dike

define the term somewhat subjectively, and in the specific context of New Mexico. For the record, here is the hierarchy of mountain taxonomy, which in everyday usage is ignored by almost everyone:

The *peak* is the basic unit (Wheeler Peak)
A group of related peaks is a *range* (the Taos Range)
A group of related ranges is a *system* (the Southern Rocky Mountains)
A group of related systems is a *chain* (the Rocky Mountains)
An extensive complex of chains is a *belt* or *cordillera.*

But this is a book about mountains, not geomorphology, and therefore I've felt free to include mountains that would be considered such only under the most liberal of definitions. Ship Rock is here, although it technically is a "diatreme." Because of their cultural importance Dowa Yalanne and Enchanted Mesa are here, even though they're more accurately labeled as mesas rather than mountains. (But with a very few exceptions, I've not included mesas.) Rabbit Ear Mountain and Mount Dora are here, simply because they're the most mountainous features in their areas. I've even included Custer Mountain! (But not Railroad Mountain.)

The issue is no less complicated when one is talking not about individual mountains but rather about mountain groups. New Mexico's largest mountain group, the Sangre de Cristos, one would think would be called a *range*, for that term connotes a large mountain group. But in fact, the accepted name is Sangre de Cristo Mountains, sharing the term with the Mud Springs Mountains, a tiny group just northwest of Williamsburg that most maps don't bother to label. On the other hand, the Fra Cristobals northeast of Elephant Butte Reservoir are really a rather modest group, but they're known as the Fra Cristobal Range, sharing the term with the Black Range to the west, one hundred miles long and one of the state's largest ranges.

In this book I've included almost all groups labeled *mountains* on the USGS 7.5-minute quadrangles. As for groups labeled *hills*, I've tended to exclude them, except when they are part of a larger group, as the Smuggler Hills are an appendage of the Animas Mountains, or when they possess mountain characteristics, such as striking relief and distinct summits. I also have included those independent *hills* with these characteristics whose natural and human histories demand that they be included. How could I omit Los Cerrillos near Santa Fe, with several large summits but more importantly the site of turquoise mines of major archaeological and historical significance?

As for whether a mountain is an independent entity or a sub-peak of a higher peak, that is equally murky. People who compile lists of mountains have established criteria; for example, "peak baggers" in Colorado intent upon ascending the state's peaks fourteen thousand feet or higher have determined that a Fourteener must be at least three hundred feet higher than neighboring summits. Similarly, hikers in the United Kingdom are fond of bagging Munros, Scottish summits three thousand feet or more above sea level. Marilyns are summits in England, Wales, and the Isle of Man with a drop all around of at least five hundred feet (150m), as listed in *The Relative Hills of Britain*. But this is a thicket of complexity into which I'm loath to enter, and as Roger Payne, executive secretary of the U.S. Board on Geographic Names, has summarized: "In the US, there is no set number defining a difference in elevation between ridges that determines whether they are separate features. There are no official definitions or terms defining or applying generic terms or required computations to any feature or entity in the naming process."

I advise not being overly concerned about any of these contradictions and inconsistencies; they occur inevitably as languages evolve, and besides, they are too firmly embedded in usage to ever be corrected. Languages and names, like mountains themselves, will ever remain rebels against categorization.

Mountain Names

No description of a mountain or a range is complete without an explanation of its name. Names are inextricably linked with the history of the people associated with the mountains. The Zuni Mountains, for example, are inseparable from the Zuni people. Similarly, the name of Vicks Peak reminds us that mountains here were a refuge for the Apache leader Victorio and his people. Often the name will describe how local people have perceived the mountain, as with Bishop Cap in the Franklin Mountains. Sometimes it reflects local folklore, as with Starvation Peak. Or the name will recall an incident that occurred at the mountain, as with Massacre Peak in the Cookes Range.

Moreover, the names often reflect the values and lifestyles of these peoples. Borrego Mesa, where rural Hispanics grazed sheep; Osha Peak, where they gathered a medicinal herb. The peoples' religious values are reflected in mountains' names as well, from references to mythology, such as Turquoise Mountain (Mount Taylor), to names commemorating Catholic saints. Only the Himalayas have an array of evocative religious names comparable to the mountains of New Mexico.

I have often stated that if you know the origin and history of the names of a region, you also know much about the culture and values of the people who created those names—and that's equally true with mountain names.

Physiographic Provinces

Southern Rocky Mountains: This is the southernmost extension of the great Rocky Mountain Chain and includes the Sangre de Cristo Mountains and the state's highest summits, formed when ancient granite cores were uplifted and exposed. In New Mexico, the Jemez Mountains, the Tusas Mountains, and the Sierra Nacimiento are included in this province, despite different geologic origins.

New Mexico's Physiographic Provinces

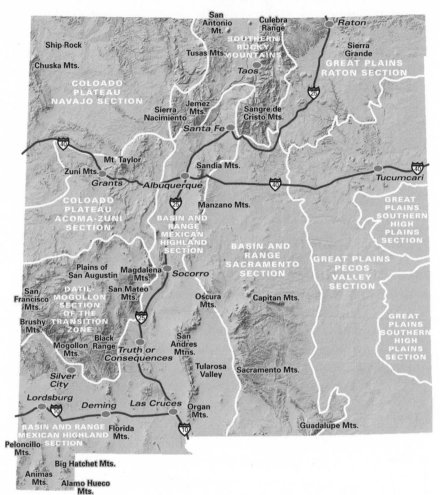

Colorado Plateau: This is a vast area of ancient, generally low-relief, sedimentary strata eroded into canyons and mesas. The province also includes numerous more recent volcanic mountains and structures, such as the Chuska Mountains, Mount Taylor, Cabezon, and Ship Rock. The Zuni Mountains, a granite-cored uplift, are the sole exception.

Great Plains: Geologically stable over millions of years and eroded to a flat plain, this province has no significant mountains, except in its northern part. There volcanism has produced a large array of volcanic mountains, including Sierra Grande, Capulin Mountain, and Eagle Tail Mountain.

Datil-Mogollon Section of the Transition Zone: As its name implies, this region is transitional between the Colorado Plateau Province to the north and the Basin and Range Province to the south. It owes its identity to relatively recent volcanism. This very mountainous region includes all the mountains of the Gila Country and the ranges around the Plains of San Augustin.

Basin and Range: Spreading of Earth's crust—and especially of the Rio Grande Rift—characterizes this highly diverse province. It includes fault-block mountains along the Rio Grande Rift, volcanic mountains such as the West Potrillos, igneous intrusion as in the Organ Mountains, and even exposure of a fossil reef, as with the Guadalupe Mountains.

Geology

Because a book about mountains almost by definition is a book about geology, I invite you herein to become acquainted with New Mexico's geology. The state's geologic history is every bit as dramatic, colorful, varied, and often violent as its human history. You'll need a strong imagination to visualize conditions and landscapes far different from those of the present, but the excitement and understanding are worth it. Especially when the insight arrives, almost as a revelation, that this drama is still being played out around us. Soon you will begin to view contemporary events, such as flash floods and rock slides, in a new, deeper context. New Mexico's geology is among the most underrated dramas of this state.

Understanding geology depends upon recognizing sequences, so I urge you to become acquainted with the broadest outlines of the geological time scale as outlined below and charted in the Geologic Glossary (see *Appendices*).

Geological Time in New Mexico

As geologists are rightly fond of reminding us, geological time is almost unimaginably long. As I write this, I'm looking out my window at southern slopes of the Sandia Mountains; what I see is what the first humans here would have seen more than eleven thousand years ago: the same canyon, even the same boulders of Sandia granite. Yet I know from geology that the boulders are ephemeral; they eroded out of the bedrock, then many tumbled downslope, often into valleys, shrinking along the way, grain by grain. Ultimately they will decompose completely, becoming one with the gravel over which I drive my car. My presence in this process is the blink of an eye.

How much harder then to grasp that the Sandia Mountains themselves are ephemeral, that before they uprose this was a broad plain, and before that were seas, and even earlier were other mountains that long ago were eroded away, just as the Sandias one day will be. And then to realize that even this is but a fraction of Earth's history. . . . As you read the following summary of how geologists have divided geologic time, please keep in mind that within these divisions, even the smaller ones, are vast amounts of time, at least on a human scale, sufficient time for major changes to have occurred. A single example: We know from continental drift that for part of its history what is now New Mexico was located on the equator. This makes the rocks at the back of my canyon seem as ephemeral as sandcastles on a beach.

Precambrian

The first of the four major geological eras is the Proterozoic, more commonly kown as the Precambrian. The name is easy to remember because of the prefix *pre-*, "before." The Precambrian is *before* almost everything—all land plants and animals, and most sea creatures as well. A few life forms existed during the Precambrian, but they were a pretty humble lot. The era began with Earth's beginning 4.6 billion years ago and lasted until 543 million years ago.

In New Mexico as in mountains everywhere, some of the oldest rocks are found atop the highest peaks. For example, New Mexico's three highest summits—13,161-foot Wheeler Peak, 13,113-foot Old Mike, and 13,102-foot South Truchas Peak, all in the Sangre de Cristo Mountains—have exposures of Precambrian rocks. In the Taos Mountains of the Sangre

Geologic Time in New Mexico

ERA	PERIOD	EPOCH	GEOLOGIC EVENTS	DOMINANT LIFE
Cenozoic	Neogene	Holocene 0.012 mya*	12,000 years ago to present Humans in New Mexico	Humans
Cenozoic	Neogene	Pleistocene 2–0.012 mya	Cooler, wetter climate; latest Jemez eruptions beginning 1.6 mya; West Potrillo Mts. erupt 0.2–.15 mya	
Cenozoic	Neogene	Pliocene 5–2 mya	Fault-block mts. continue to rise; eruptions in Raton-Clayton area; Mt. Taylor 2.5 mya	Mammals
Cenozoic	Neogene	Miocene 23–5 mya	Sandia-Manzano uplift 20 mya; first Jemez eruptions 18 mya; uplift of Guadalupe Reef 10 mya; Raton-Clayton volcanic field begins 8 mya	
Cenozoic	Paleogene	Oligocene 34–23 mya	Rio Grande Rift and Basin and Range faulting begin; widespread volcanism in southern New Mexico; laccoliths of Los Cerrillos, Ortiz, San Pedro, South Mt., and Capitan Mts. 30 mya; Plains of San Agustin volcanism 26–24 mya; Colorado Plateau diatremes 25–30 mya	
Cenozoic	Paleogene	Eocene 56–34 mya	Sedimentary basins due to compression continue; Datil-Mogollon volcanism begins 40 mya	
Cenozoic	Paleogene	Paleocene 65–56 mya	Late Laramide volcanism; hills and stream-crossed lowlands of compressional basins	
Mesozoic	Cretaceous 145–65 mya		Mass extinction 65 mya; Laramide orogeny; Rocky Mts. arise again; broad seaway across midcontinent	Dinosaurs
Mesozoic	Jurassic 200–145 mya		Lakes and streams across northern New Mexico; hills to south; seaway in Bootheel	
Mesozoic	Triassic 251–200 mya		Hills in northern and southern New Mexico; streams and lakes in between	
Paleozoic	Permian 299–251 mya		Permian Extinction 251 mya; Guadalupe reef forms 260 mya	Amphibians
Paleozoic	Pennsylvanian 323–299 mya		Ancestral Rockies, marine limestone now exposed atop Sandias; Zuni Mts. and part of Fort Defiance uplift; Mts. and basins criss-cross region later to become Sangre de Cristos	
Paleozoic	Mississippian 359–323 mya		Biological mounds in limestone of southern New Mexico	Fish
Paleozoic	Devonian 413–359 mya		Seas continue pattern of advancing and retreating	
Paleozoic	Silurian 444–416 mya		Seas and land similar to Ordovician; earliest land plants	
Paleozoic	Ordovician 488–444 mya		Seas and land similar to Cambrian	Invertebrates
Paleozoic	Cambrian 542–488 mya		Seas cover southern New Mexico; arch of land in central and northern New Mexico	
Proterozoic	Precambrian 2,500–542 mya		Evolution of simple life forms; metamorphic complexes in cores of later mt. ranges	Simple primitive life forms
Proterozoic	Archean 4,600–2,500 mya		Earliest preserved rocks and life forms	

*mya=million years ago

© 2006 K. Sparkes

Scale of geologic time in millions of years

65
186
291
1500+

de Cristos are the state's oldest rocks, highly metamorphosed volcanic rocks approximately 1.8 billion years old, more than a third as old as Earth itself. For most of their history, these ancient metamorphic rocks, formed by heat and pressure from still earlier rocks, lay buried; then about 20 million years ago they were exposed by the uplift that also created the Rio Grande Rift.

Precambrian rocks are found, north to south, in the Sangre de Cristo Mountains, including the Taos Mountains, Cimarron Range, Picuris Mountains, Rincon Mountains, and Santa Fe Mountains; Tusas Mountains; Sierra Nacimiento; Chuska Mountains; Sandia, Manzanita, Manzano, and Los Pinos Mountains; Pedernal Hills; Zuni Mountains; Sierra Ladrones; Caballo Mountains; Oscura and San Andres Mountains; Burro Mountains; Franklin Mountains; Florida Mountains; and Animas Mountains. Precambrian rocks, especially granite, form the core of many other ranges. These Precambrian rocks sometimes are igneous rocks such as granite; often they are metamorphic rocks such as quartzite, gneiss, and schist, whose tiny, toughly compacted grains once were sand on the beaches of ancient oceans. In his award-winning book about geology, *Basin and Range,* John McFee said: "If by some fiat I had to restrict all this writing to one sentence, this is the one I would choose: The summit of Mountain Everest is marine limestone." Atop New Mexico's highest mountains are memories of ancient oceans.

Paleozoic Era

Next in the four-era sequence is the Paleozoic, also easy to remember because of *paleo-*, "old." Life really blossomed here, but it was a long, long time ago. The Paleozoic Era is made up of several periods.

Cambrian Period: 542–488 million years ago southern New Mexico was covered by a shallow sea, extending from the south, from whose beaches sandstone sediments were deposited. Among the most important of these formations is the Bliss Formation, a complex of sandstone-rich strata visible today on the eastern escarpment of the San Andres Mountains and the western escarpment of the Caballo Mountains as a darkish band just above underlying, pinkish, Precambrian granite. Almost all the mountains of southern and southwestern New Mexico have outcrops of Cambrian and early Ordovician rocks.

Ordovician Period: 488–444 million years ago. The Bliss Formation spanned the Cambrian-Ordovician boundary as the southern sea continued to cover the region. The limestone making up the El Paso Formation, which often succeeds the Bliss sandstones, was deposited in the Ordovician sea. This sea teemed with vertebrate creatures, but later erosion removed from southern New Mexico all the rocks bearing their fossils. Erosion also left New Mexico with no strata from the succeeding Silurian Period, 444–416 million years ago.

By the Devonian Period, 416–359 million years ago, the seas had receded, and most of southern New Mexico was a muddy, swampy lowland, reminiscent of today's coastal mangrove swamps. Again, erosion erased all but the most recent 20 million years of Devonian strata, which crop out in most of the mountain ranges of south-central and southwestern New Mexico, from the Sacramentos on the east to the Peloncillos on the west. These are mostly marine shales and siltstones, with brachiopods as common fossils.

Better represented are similar formations from the Mississippian Period, 359–318 million years ago. Mississippian marine shales were laid down and still exist in the Sangre de Cristo Mountains northeast of Santa Fe and in the Sierra Nacimiento east of Cuba. Most of the Mississippian deposits in northern New Mexico were eroded later in the period as the sea retreated. Southern New Mexico, however, was more uniformly submerged, and massive fossil-bearing limestone formations were laid down, then partially eroded by Pennsylvanian time. Nonetheless, Mississippian rocks are now exposed in the Sacramento, San Andres, and Big Hatchet Mountains.

This pattern of advancing and retreating shallow seas and adjacent lowlands ceased during the Pennsylvanian Period, 318–299 million years ago. This was a time of mountain building, of deep basins bounded by extensive uplifts. The Ancestral Rockies arose, and while the Uncompahgre Range in Colorado and Utah, north of present-day New Mexico, was eroded away long ago, the seemingly insignificant Pedernal Hills south of Clines Corners stubbornly recall the once-grander Pedernal Uplift that stretched almost to Texas.

The land continued to rise during the Permian Period, 299–251 million years ago. Then as now, mountains shed debris from their slopes, filling basins and valleys, and the eroded sediments from these ancient mountains today make up many of the reddish, sedimentary layers found in the Sangre de Cristos. More significantly, off the coastline of the Permian seas in southern New Mexico, around what geologists have labeled the Delaware Basin, began growing the giant Capitan and Goat Seep reefs that one day would be uplifted to become the limestone Guadalupe Mountains. As Barry S. Kues and Katherine A. Giles wrote in *The Geology of New Mexico: a Geologic History*, "The Capitan Reef is the largest, best-preserved, most accessible, and most intensively studied Paleozoic reef structure in the world."

New Mexico at the end of the Paleozoic was a place fecund with plant and animal life, evolving explosively as in today's rain forests.

And then it ended—abruptly.

Not as well-known as the dinosaur extinction at the end of the Mesozoic Era, the Permian Extinction was the greatest in the history of life on Earth; seventy percent of all land species—and eighty-five percent of all marine species—ended. Scientists still debate the cause.

Mesozoic Era

The Mesozoic Era dawned with a new assemblage of plants and animals. Most of present-day New Mexico was above water, with a landscape of extensive plains except for a northwest-trending mountain range in the far, north-central region. This was the era of giant reptiles, whose fossils have international recognition.

First of the Mesozoic Era's three periods was the Triassic Period, lasting from 251 to 200 million years ago. With most of New Mexico above sea level, any sediments were lake, stream, or wind deposits and thus were possible repositories for the fossils of land animals, such as amphibians, crocodilians, and early dinosaurs. Upper Triassic outcrops occur throughout northern New Mexico, often on the flanks of mountain ranges. The Chinle Formation, laid down during the Jurassic Period, is well known for sometimes containing the fossilized trunks of ancient trees.

During the Jurassic Period, 200–145 million years ago, New Mexico was again primarily above the ocean, a landscape of featureless lowlands. Lakes, streams, and swamps teemed with life, but the Jurassic is inevitably linked with dinosaurs, including New Mexico's most famous dinosaurs— Coelophysus, the "state dinosaur," and Seismosaurus, which at the time of its discovery in the 1980s by Arthur Loy and Jan Cummings of Albuquerque, was the world's largest dinosaur, with a length of up to 170 feet and a weight of up to 30 tons. The Morrison Formation is the most widely known, vertebrate fossil-bearing stratum of the Jurassic. But except for highlands in southern New Mexico, few mountains existed in the Jurassic.

Even into the Cretaceous Period, 145–65 million years ago, at the end of the Mesozoic, the story was one of shallow seas advancing or retreating over lowlands, with volcanoes erupting in the southwest and northwest quadrants. In these vast lowlands, New Mexico's coal deposits were formed. It was only toward the end of the Cretaceous that the changes occurred, when a continent-wide uplift, the Laramide Orogeny, wracked western North America. The seas departed and haven't returned. Major mountains arose, including the Rocky Mountains and their southernmost range, the Sangre de Cristo Mountains, and such smaller ranges as the Sierra Nacimiento. The Laramide Orogeny continued into the next era, the Cenozoic.

Cenozoic Era

A note on terminology: In 2003, the International Commission on Stratigraphy (ICS), the committee of the International Union of Geological Sciences that is responsible for setting geologic names and dates, approved a major change in Cenozoic labeling: the terms Tertiary (65–1.8 million years ago) and Quaternary (1.8 million years ago to present) no longer will be used. Replacing them will be Paleogene (65–23 million years ago) and Neogene (23 million years ago to present). The terms Tertiary and Quaternary have been used by geologists for a long time, and doubtless will continue to be used and recognized. In this book, however, I have chosen to use a term like mid-Cenozoic rather than Tertiary to label the time when so many important geological events in New Mexico occurred.

Almost all of New Mexico's present mountains were created during the Cenozoic Era. Geologists divide the era into the Paleogene and Neogene Periods, further divided into the following epochs, from oldest to most recent: Paleocene, Eocene, Oligocene, Miocene, Pliocene, Pleistocene, and Holocene. The Pleistocene began about 1.8 million years ago, while the Holocene Epoch began a mere 11,000 years ago and is still continuing.

The Laramide Orogeny, sometimes called the Laramide Revolution, has been the defining event of the Cenozoic. And a "revolution" it has been indeed, for it overthrew the *ancien régime* of shallow seas and featureless lowlands that for millions of years had dominated New Mexico.

It has not been a peaceful revolution—and we have no reason to believe it is over. During the Laramide Orogeny, mountain-building cracked Earth's surface in New Mexico; wedges of the crust slid vertically against each other along fault lines.

The impact of all this spread far beyond central New Mexico. Stress on

Earth's crust allowed magma chambers to ascend, and when they emerged at the surface, volcanoes erupted. Except for the Caprock of southeastern New Mexico, no portion of the state escaped major volcanism. The list of volcanic formations from the Cenozoic includes most of New Mexico's well-known individual mountains: Ship Rock, San Antonio Mountain, Sierra Grande, Capulin Mountain, Sierra Blanca, and Mount Taylor. Volcanic activity also spawned most of New Mexico's mountain ranges, including the Jemez, Magdalena, San Mateo, White, and Chuska Mountains, as well as the Black Range and almost every mountain group in New Mexico's entire southwest quarter.

Then in the southwest, beginning about 36 million years ago, volcanoes of a scale far beyond anything in human memory exploded again and again, blanketing the region with ash, lava, and mudflows, and pushing older formations upward. Using space satellites we can still see the rough outlines of a gargantuan caldera in the Gila Country. This region now is called the Datil-Mogollon Volcanic Field, and its numerous mountains testify to widespread and complex volcanism: the Magdalena, San Mateo, Tularosa, Elk, Brushy, Kelly, Jerky, Mogollon, and Diablo Mountains, as well as the Black Range and portions of the Burro Mountains. Hikers in these widely separated ranges will notice strong similarities in the rocks and formations.

Volcanic uplift and eruptions also created the mountains farther south, from the West Potrillos to the Floridas to the Big and Little Hatchets to the Pyramids and Peloncillos, among others. The hot springs throughout the region are distant echoes of the fiery cataclysms here.

Following soon after the southwestern volcanism, volcanoes in other regions were erupting to create Sierra Blanca and the Sacramento Mountains in south-central New Mexico and the Mount Taylor complex in western New Mexico. Volcanic activity also created the Jemez Mountains, whose central Valles Caldera is conspicuous on images taken from space. Farther north, along the spreading Rio Grande Rift, volcanism built the hulking shield volcanoes of San Antonio Mountain, Cerro del Aire, Cerro de la Olla, and others north of Taos. In northeastern New Mexico, the Raton-Clayton and Ocate (Mora) volcanic fields became active, spawning scores of volcanic cones. Sierra Grande, a huge shield volcano, and Capulin Mountain, a shapely cinder cone, date from this time.

Concurrent with the upheaval that triggered the volcanoes, ancient, long-buried, granitic formations were uplifted. Erosion stripped off their mantle of sedimentary rocks and exposed the resistant igneous rock beneath as mountains. The Capitans, Zunis, and Sierra Nacimiento are examples.

To the present

Bringing New Mexico geological history to the present are the most recent two epochs of the Neogene Period. The Pleistocene began about 1.8 million years ago and has become well known as the time of the great ice ages. New Mexico never even came close to experiencing the great continental ice caps that hunkered over northern North America, but mountain glaciers did exist here, sculpting U-shaped valleys among the high peaks of the Sangre de Cristos and scooping bowl-shaped cirques on mountain slopes as far south as Sierra Blanca. On the sides of some ranges, such as the Capitan and the San Mateo Mountains, ice accumulated in talus slopes and produced "rock glaciers." Also during the Pleistocene, melt water from large snow packs shaped the landscape through erosion.

Volcanism persisted during the Pleistocene, creating the lava flows around Grants and Carrizozo. It was during the Pleistocene that the final Jemez Mountain eruptions occurred.

The last epoch is called the Holocene, or Recent. It began with the end of the Pleistocene ice ages about eleven thousand years ago and continues today. Our time perspective is too brief for us to know how geologists millions of years hence will classify our time. Perhaps they'll say it was just a minor interlude between ice ages; we are still in the Pleistocene but just don't know it. Perhaps they'll note that volcanism continued, and that twenty-first century humans experienced a transitory quiescent period. After all, Capulin Volcano erupted a mere fifty-six thousand years ago, and some of the state's most recent lava flows are as little as three thousand years old (at El Malpais National Monument). And no doubt the uplift-subsidence along the Rio Grande Rift is continuing. Perhaps those future geologists will come here to behold one of Earth's greatest natural wonders: thirty thousand-foot mountain ranges paralleling a fifteen thousand-foot-deep gorge.

But for us, it is enough to know that we are in the midst of one of Earth's greatest geological dramas, and that the scripts from previous acts are everywhere for us to read.

Young volcanoes on Albuquerque's West Mesa,
flanking the Rio Grande Rift

Mountain building in New Mexico

Five geological processes—folding, uplift of fault blocks, igneous intrusion, erosion, and volcanism—account for most mountain-building, and all are represented in New Mexico. Actually, we have a sixth here in New Mexico: none of the above quite encompasses the formation of the Guadalupe Mountains, i.e., uplift and exposure of a fossil reef. There are few places where this occurs, but in New Mexico it is significant.

Examples of each type:

1. Folding: Mud Springs Mountains, Robledos, Franklins
2. Fault block: Sandias, Manzanos, Caballos, San Andreses, Oscuras
3. Igneous intrusion: Capitans, Zunis, Organs, Ortiz/San Pedro/South Mountains, Peloncillos, Sangre de Cristos, Sierra Nacimiento
4. Volcanism: all the mountains of the Datil-Mogollon Volcanic Field, the Jemez Mountains and individual volcanic formations along the Rio Grande Rift such as San Antonio Mountain, Sierra Grande and

related northeastern mountains, West Potrillos, Floridas, Ship Rock, and the other volcanic plugs

5. Erosional: Chuska Mountains, Huerfano Mountain, Starvation Peak, Tucumcari Mountain

6. Fossil reef formation: Guadalupe Mountains (Eddy County).

Rio Grande Rift

Much of New Mexico's landscape today has its origins in events that began 30 million years ago and a thousand miles away, at the interface between the Pacific and North American plates. These are among the global-scale pieces of the Earth's crust that are moving, albeit very slowly by human standards, over the Earth's surface. The interaction of these two plates caused the mid-section of the North American continent to arch upward, thinning and stretching Earth's crust in western North America. This stretching and thinning of the crust has allowed magma deep beneath the surface to rise. Eventually, this stretching and thinning of the crust caused it to split, manifesting in a major break in the crust that runs all the way from Central Colorado near Leadville south through New Mexico and into the Mexican state of Chihuahua.

Often the rising magma bodies emerged at the surface as lava flows and volcanoes. Many of New Mexico's volcanic mountains were created then, from Sierra Grande in the northeast, to San Antonio Mountain along the Rio Grande north of Taos, to the huge volcano of Sierra Blanca, and to the numerous volcanic mountains of the Datil-Mogollon Volcanic Field and the Basin and Range Province of southwestern New Mexico.

Then, beginning about 17 million years ago, the second phase of the rifting began. As the crustal extension continued, the long sliver of land between the two parts dropped downward between parallel north-south faults, forming a *rift*. In some places along the Rio Grande Rift the land has subsided as much as twenty-six thousand feet below its pre-rift elevation.

Meanwhile, the land along the rift's margins was tilted upward. Imagine a sidewalk segment broken by a root pushing upward. The sections facing the break—the rift—will be steep and jagged, while the rest of the pavement block will slope gently away. This is why mountains along the rift, such as the Sandia and Manzano Mountains, have steep slopes facing the Rio Grande Valley but gentler slopes facing away.

Throughout this upheaval, magma has continued to rise, and along the

The gorge of the Rio Grande Rift, on the Rio Grande near Questa

Rio Grande Valley are numerous relatively recent volcanic features, from Mount Taylor to the volcanoes on Albuquerque's West Mesa, as well as some of North America's most recent lava flows. That much of this lava is basaltic suggests that the faults along the Rift are very deep, reaching down to the mantle, which is the source of most basaltic magma.

Also, the fracturing of the rocks along the Rift, along with magma upwelling, has allowed penetration by superheated, mineral-rich solutions, creating the mineral deposits upon which much of New Mexico's mining has been based.

The Rift is called the Rio Grande Rift because of the river flowing through it, but this is not an ordinary river valley. It was not cut by the river and does not branch upstream, like most river valleys. Rather, the river simply took advantage of the valley created by the rifting. In fact, early in its history the rift consisted of four basins that, when filled with water, were huge lakes connected by streams. Eventually erosion erased the barriers between the lakes, and a continuous stream flowed through the channel, the Rio Grande of today.

The stretching and thinning that created the Rio Grande Rift, along with the upwelling of magma along the Rift's margins, is continuing today. The Sandia Mountains are a mere 10 million years old and still growing. Beneath the town of Socorro a magma chamber is rising that is expected to reach the surface in seven hundred twenty-six thousand years, perhaps creating New Mexico's newest volcano.

The Rio Grande Rift is the defining feature of New Mexico's present landscape. It is the hinge upon which much of the state's present geological history has turned.

The Volcano State

From all over the world, geologists interested in volcanoes come to New Mexico. Dr. Jayne Aubele and Dr. Larry Crumpler, geologists with the NM Museum of Natural History and Science in Albuquerque, say a strong case can be made for New Mexico being the nation's Volcano State: "New Mexico has one of the greatest concentrations of young, well-exposed, and uneroded volcanoes on the continent. And as a bonus, it is also the Rift Valley state; it has one of only four or five big continental rifts in the world, East Africa being one of the other ones. The fact is, New Mexico is one of the best places to study the natural history of volcanoes."

In addition to having one of the largest numbers of volcanoes, New Mexico's volcanoes also have one of the largest spans of ages, diversity of types, and range of preservation, as well as some of North America's best type examples. When asked the common question, "Will New Mexico see another eruption?" Aubele and Crumpler answer, "Probably." When? "Geologically soon." Don't hold your breath, but you might want to keep your eye on the Socorro area, as deep underground there—and expanding upward—is one of only three large, mid-crustal, active magma bodies on the continent, the others being in Long Valley, CA, and Yellowstone, WY.

Ecology

Ecologists have said that New Mexico includes all the life zones in North America—except tropical. (During the state's "monsoon season," some residents might even include that!) Certainly biological and topographic diversity are part of the state's appeal among outdoor recreationists. Not only can they get out somewhere in the state at any time of year, they also have an extraordinarily broad array of ecosystems to explore; in the same week they can hear pikas barking among rocks above timberline in the Sangre de Cristo Mountains, and also be on the alert for Gila monsters in the Peloncillo Mountains.

The major determining factors behind this diversity are latitude and elevation. New Mexico shares its southern border with Mexico and Texas, in the Chihuahuan Desert, but its northern border is with Colorado, in the Southern Rocky Mountains. Elevations in the state range from 2,817 feet near the Texas border south of Carlsbad to 13,161 feet atop Wheeler Peak in the Sangre de Cristo Mountains northeast of Taos, a vertical relief of more than 10,000 feet. The occurrence of specific plants and animals within this enormous range depends on other factors as well, such as available moisture, temperature, habitat, soils, and so forth, with thousands of micro-ecosystems existing throughout the state. Because of these variations, New Mexico can be divided into several major life zones that correlate primarily with elevation; temperatures decrease by 3.6°F for each thousand feet gained in elevation, while capture of moisture increases. Outdoor recreationists should become familiar with these life zones, not only because an awareness of them enhances their experience but also because each has its unique characteristics and challenges. Though classification systems and labels vary widely among naturalists, and the actual situation is much more complex than what is presented here, the following classification has been widely used.

Lower Sonoran. *Note:* The terms Lower Sonoran and Upper Sonoran are used by ecologists to refer to specific life zones and do *not* refer to the Sonoran Desert itself. The Sonoran Desert is not found in New Mexico; the primary desert here is the Chihuahuan Desert (the Great Basin Desert extends into the northwestern part of the state), whose plants are classified as Lower or Upper Sonoran. The Lower Sonoran life zone occurs below four thousand five hundred feet in New Mexico. Temperatures are hot for much of the year and evaporation is high. Characteristic plants are mesquite, acacia, creosote bush, yucca, agave, ocotillo, fourwing saltbush, cholla and numerous other cacti, with cottonwood, box elder, bigtooth maple, oak, and other deciduous trees in isolated riparian environments. Again, these plants of the Chihuahuan Desert are typical of the Lower Sonoran life zone.

Life Zones on Mountain Slopes

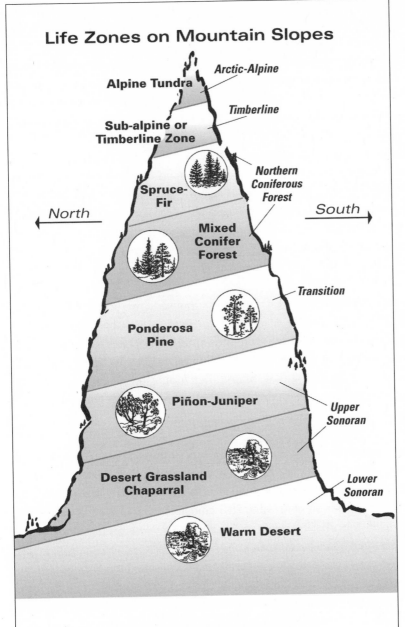

Each tree species has a range of altitude.
The upper limit is the altitude where it becomes too cold for that
tree to grow. The lower limit is where it is too dry.

Upper Sonoran. Within this zone, generally occurring to seven thousand five hundred feet, summers are hot, winters somewhat mild. Precipitation is modest, and evaporation is still high. This is the zone of the piñon-juniper forest, New Mexico's most widespread vegetation type. Other characteristic plants are mountain mahogany, alligator juniper, Chihuahuan pine, cholla, Apache plume, Emory and other oaks, and, near water, boxelder, tamarisk, and, of course, cottonwood. There is considerable overlap between the Lower and Upper Sonoran life zones.

Transition. Within this zone, seven thousand five hundred feet to eight thousand two hundred feet, summers are mild, winters cold, often with substantial snowfall. Here ponderosa pines replace piñons and junipers, and Gambel oaks become common. Native pines such as limber pine and Mexican white pine also occur here. Other typical plants are Rocky Mountain maple and New Mexico locust.

Northern Coniferous Forest. From eight thousand two hundred feet to timberline (ten thousand to eleven thousand five hundred feet in New Mexico) summers are cool, winters cold; mountains with these elevations catch significant moisture. This zone is dominated by Douglas fir, white fir, subalpine fir, Engelmann spruce, blue spruce, alpine juniper, limber pine and, occasionally in the north, bristlecone pine. Aspens are common in disturbed or transition areas.

Arctic-Alpine. Here above timberline temperatures are cool to cold, even in summer; evaporation is high because of wind, elevation, and exposure; and while considerable snow falls, most is swept away by high winds. The dominant vegetation here is a tough, windswept mat of low, hardy grasses, sedges, and shrubs. Travelers in this zone should be prepared for severe weather at all times of the year.

The above classification is greatly simplified, and the elevation delineations vary by latitude; for example, Upper Sonoran plants in the southern part of the state will be found at higher elevations than in the northern part. Also, plant distribution is affected by whether the slope on which they grow is north- or south-facing. Don't get hung up on labels or definitions, just get to know the plants.

Ecosystems and Life Zones

Within each life zone are numerous ecosystems created by localized conditions of elevation, aspect, water, soil, and more. The ecosystems described in each text block are based upon the *New Mexico Land Cover Map: New Mexico Gap Analysis Project.* This was a U.S. Department of the Interior initiative that in New Mexico involved numerous federal agencies and university programs. They cooperated in analyzing satellite images, as well as field checking, to produce a master map of the dominant vegetation on the land. The land cover classes were developed by Esteban Muldavin, ecologist with the NM Natural Heritage Program.

Because most readers are more familiar with actual plants, such as desert sage and ponderosa pine, rather than terms such as Great Basin Monophyllous Desert Scrub, I've used the plants to define the ecosystem. Readers wishing the formal ecological classes should consult the Gap Analysis Project at the University of New Mexico.

Climate and Mountain Morphology

New Mexico's arid climate and geographical location have had a profound effect on the region's mountains. Ninety percent of New Mexico receives less than twenty inches of precipitation annually, while twenty to thirty percent receives less than ten inches. The highest averages, probably not greater than thirty-five to forty inches, are limited to extremely small, high-elevation areas in the northern mountains.

This means the mountains in dry areas are sparsely vegetated, and their thin soils are fragile and exposed. At the same time, the pattern of summer monsoon thunderstorms means that individual rains often have great erosive power and carry enormous sediment loads. To witness a flash flood here is to see years of erosion occur in an hour. Thus, mountains here display an angularity that rarely survives in wetter areas, where erosion is more constant but more gentle. Runoff from these brief but powerful storms create alluvial fans around the base of mountains; these are especially conspicuous in the Basin and Range mountains of southern New Mexico.

Climate has also affected New Mexico's mountain landscape through glaciation, or rather its absence. New Mexico was far from the continental ice sheets of the Pleistocene, but valley glaciers existed here. They were common in the Sangre de Cristo Mountains, as they were in the rest of the

Rockies. But most mountains south of the Sangre de Cristos never experienced glaciation; Sierra Blanca, the southernmost mountain in the United States to have had a glacier during the recent Ice Age, is the exception.

Mechanical weathering of rocks in arid regions is often abetted by broad temperature variations. Even allowing for the state's ten thousand feet of vertical relief, New Mexico has far greater extremes of climate than most places. Its lowest recorded temperature (-50°F), recorded on February 1, 1951, at Llaves, north of Cuba, elevation 7,350 feet, is colder than the record lows in New Hampshire, Maine, Washington, California, and thirty other states. The state's highest temperature, recorded at 3,418 feet near Carlsbad on June 27, 1994, is 122°F, exceeded only by higher temperatures recorded in California, Arizona, and Nevada.

Mountains and New Mexico's Peoples

When the first humans arrived in what is now New Mexico, at a time in the past whose dates still are hotly debated but no later than around 13,500 years ago, they would have beheld a land of mountains. From Raton Pass or from the northeastern plains they'd have seen huge, extinct volcanoes, and to the west the great range of the Sangre de Cristo Mountains. Had they entered the region by following the Rio Grande south, they'd have passed through a corridor bordered by the Sangre de Cristos to the east and the Tusas Mountains and Jemez Mountains to the west. Had they entered via the Colorado Plateau, they'd have seen volcanic formations such as Mount Taylor and ranges such as the Chuska and Zuni Mountains. And as they migrated southward, they never would have been out of sight of other mountains.

Thus, as the dominant landform throughout most of New Mexico, mountains have been a pervasive—and in some instances, formative— influence on the cultures that have evolved here, and mountains have been a part of the daily reality of all peoples living here, except for those on the eastern and southeastern plains. To be sure, the oldest archaeological sites are on the plains, where Paleo-Indians hunted the Pleistocene megafauna. But we know from their artifacts that Paleo-Indians traveled throughout the state, and when the climate became warmer and drier about seven thousand years ago, and most of the large mammals became extinct, hunter-gatherer Indian peoples increasingly altered their food gathering to emphasize the more diverse mountain environments.

New Mexico's mountains provided hunter-gatherers with what they needed most: food, water, and shelter. And even when some Indian peoples began moving toward agriculture after approximately 3,500 BP, they never completely severed their dependence upon resources obtained in mountains. Detailed chemical analyses of large timbers in the ruined Great Houses of Chaco Canyon reveal their source to be the Chuska Mountains and Mount Taylor. Chacoans also traveled to the Jemez Mountains to quarry obsidian. Native peoples from throughout New Mexico traded in turquoise mined in the Little Hatchet Mountains, the Big Burro Mountains, and in Los Cerrillos near Santa Fe at the famous mine at Mount Chalchihuitl.

Mountains have influenced the development of Indians' mythological and religious beliefs and served as territorial markers. Four sacred mountains define the Navajo universe: the San Francisco Peaks in Arizona, the mountains of the West; the La Plata Mountains in southern Colorado, the mountains of the North; Sierra Blanca, in Colorado's Sangre de Cristo Mountains, the mountain of the East; and Mount Taylor, the mountain of the South. The soul center of the old Navajo homeland, Dinétah, is an obscure (to non-Navajos) mountain bump called Gobernador Knob. Sierra Blanca in southern New Mexico is sacred to the Mescalero Apaches. As William deBuys has observed of the Puebloan peoples:

> The mountains are active forces in the spiritual landscape of the Pueblos. Elsie Clews Parsons even suggests in *Pueblo Indian Religion* that the mountains were no less than deities themselves, since they were represented in effigy in numerous rituals that take place within the pueblo. Whatever rank one may ascribe to them, the sacredness of the mountains signifies not the rarification of the life force but the ubiquity and abundance of it. (*Enchantment and Exploitation: the Life and Hard Times of a New Mexico Mountain Range*)

Thus, contemporary Indians continue to maintain shrines on many mountain summits, and Navajos, Mescaleros, and Acomas restrict non-Indian access to their sacred mountains of Ship Rock, Sierra Blanca, and Enchanted Mesa (Katzimo).

When the Spanish arrived in 1540, they, like their native predecessors, depended upon the mountains for sustenance in the form of game, fish,

Petroglyphs on volcanic rocks on Albuquerque's West Mesa

timber, fuel wood, medicinal and edible plants, and perhaps most important, water. By capturing and storing winter snow, mountains are the source
of all of New Mexico's perennial streams and rivers, and thus make irrigation possible. The Spanish also used mountains for grazing, taking cattle
and sheep into high meadows in the summer, a traditional land use continuing today, though not without some controversy, especially when linked
with the issue of land grants and use of commons lands.

The spiritual relationship of the Spanish with mountains, however, was
markedly different from that of the Native Americans. Rather than being
seen as spiritual entities themselves, mountains became identified with
saints and other icons of the Catholic faith. Mountain names here reflect
this: a pantheon of saints—San Agustín, San Andres, San Antonio, San
Diego, San Francisco, San Luis, and San Mateo—as well as others such as

Animas, Sacramento, Cerrito del Padre, Sierra del Cristo Rey, and greatest of all, Sangre de Cristo. I once wrote a book, *Mountain Names,* about the names of the world's major mountains and ranges. I was impressed by the beauty of the mountain names in the Himalayas, names derived from the Hindu religion. Then I realized the only place on Earth I could recall with a similar array of poetic and beautiful religious mountain names was New Mexico. Hispanics, like Native peoples, often construct shrines on the high points of hills and mountains; examples include El Cerro de Tomé, Hermit Peak (*El Solitario*), Cerro Pedernal, and El Porvenir.

The Spanish also mined wherever they found promising ore, but Indian dangers and primitive technology kept most operations small. Large-scale mining in New Mexico's mountains did not begin until the significant influx of Americans in the mid-1800s. Hispanics did, however, graze large numbers of cattle and especially sheep in New Mexico's mountains, with major impacts on vegetation and other aspects of mountain ecology.

By the late 1800s, deforestation had seriously degraded many watersheds—for example, logging left the Zuni Mountains almost completely denuded—while market hunting had decimated mountain wildlife, not only in New Mexico but throughout the nation. This led to the establishment of the first Forest Reserves in 1891. In New Mexico the National Forests, especially in the north, came to include lands formerly within Spanish and Mexican land grants, igniting a perennially smoldering controversy. Today, most of the state's forested mountains are within one of the state's five National Forests. Other mountain lands are owned and managed by the U.S. Bureau of Land Management.

Concomitant with the establishment of the National Forests was the growing concept of wilderness preservation. In 1924, the Gila Wilderness became the first area anywhere in the world to be set aside solely to preserve its natural qualities. The Pecos country had been set aside even earlier, in 1898, for watershed protection. The San Pedro Parks in the Nacimiento Mountains was declared a Primitive Area in 1931. The Pecos country received similar designation in 1933.

Yet for decades New Mexico and the nation lacked a comprehensive system of wilderness preservation, and it was not until 1964, despite decades of false starts, that Congress passed the Wilderness Act. It included this definition: "A wilderness, in contrast with those areas where man and his own works dominate the landscape, is hereby recognized as an area where the earth and its community of life are untrammeled by man, where man himself is a visitor who does not remain." That clearly described many of New Mexico's mountains.

Yet while the San Pedro Parks and the White Mountains received wilderness designation under this act, it took sixteen years of further study and sustained involvement by citizen wilderness groups to produce the New Mexico Wilderness Act of 1980. This created several more wilderness areas, including Apache Kid, Withington, Capitan Mountains, Dome, Latir Peak, and Cruces Basin. It also left Forest Service Wilderness Study Areas in place in the Peloncillo and Sangre de Cristo Mountains.

At the same time the U.S. Bureau of Land Management (BLM), responding to the mandate of the 1976 Federal Land Policy Management Act, was inventorying its own extensive holdings in New Mexico for possible wilderness designation. This resulted in numerous BLM Wilderness Study Areas, most associated with mountains in southern New Mexico. And again, a group of private citizens—the NM Wilderness Coalition and later the NM Wilderness Alliance—conducted their own inventory and promoted their own recommendations. As of this writing, no comprehensive BLM wilderness bill has been introduced into Congress. In the meantime the WSAs are mandated to be managed to preserve their wilderness characteristics. One area, the BLM Ojito Wilderness, was established in 2005.

Today, New Mexico's mountains serve different human needs than when people first arrived here thousands of years ago. No longer do we seek obsidian for projectile points, and while many traditional uses have survived—hunting and fishing, gathering of edible and medicinal plants—most of what we seek from mountains is more subtle but no less real. These include opportunities to enhance our physical and mental health, to experience alternatives to our urban lifestyles, to reconnect with nature, and to surrender to the profound fascination members of our species have always had for mountains.

Mining and New Mexico's Mountains

Mountain building and mineralization often are linked, and thus so are mountains and mining. In what is now New Mexico, mining dates back to the earliest human habitation as high-quality obsidian quarried in the Jemez Mountains has been found at Paleo-Indian sites through the West.

About 1,100 years ago, Indians of the Southwest also began mining turquoise. New Mexico's most famous turquoise mine is Mount Chalchihuitl, in Los Cerrillos south of Santa Fe, but prehistoric turquoise mines also existed in the Little Hatchet, Burro, and Jarilla Mountains.

Mineral extraction in New Mexico remained modest, at best, even after the arrival of Spaniards in 1540, despite their passion for mineral wealth and prospecting. Indeed, one could claim that Coronado became the first European prospector when his expedition arrived here expecting to find gold. (They also expected the Indians to have previously mined and shaped it, as the Aztecs and Incas had obligingly done.) The Spaniards were profoundly disappointed to find only turquoise in the Indian villages, but like treasure hunters everywhere, the Spaniards simply kept looking. Nor did the apparent paucity of gold and silver deposits deter the proliferation of legends about fabulous Spanish mines hidden by Indians after the Pueblo Revolt in 1680.

Actually, in 1828 a New Mexican did make a rich gold strike. That year José Francisco Ortiz, a sheepherder in the Ortiz Mountains, discovered placer gold there. Soon, prospectors discovered vein deposits. Although the strike occurred in what was then Mexico, it since has been hailed as the first major American gold strike west of the Mississippi. In the mountains north of Silver City, several *arrastres*, crude, animal-driven, ore-milling devices, have been found, but their output was minor, and the most significant Spanish mining in southern New Mexico occurred at Santa Rita, where rich copper deposits discovered in 1798 continue to be mined today.

The localized and generally insignificant mining during the Spanish and Mexican periods changed dramatically, like so much else, with the Americanization of New Mexico beginning in 1846. Prospectors, many of them former Army soldiers, swarmed over New Mexico's mountains, despite fierce resistance by resident Indians. In August 1878, George Daly discovered in the foothills of the Black Range one of the world's richest silver ore bodies. Called the Bridal Chamber, it yielded 2.5 million ounces of silver chloride, some so pure that miners used saws to extract it. On the day of his discovery, Daly was killed by Apaches.

But by the 1890s, the Apaches were gone, and in mountains throughout New Mexico mining camps sprang up including Elizabethtown in the Cimarron Range, Rosedale in the San Mateo Mountains, Bland in the Jemez Mountains, Chloride and Winston in the Black Range, Mogollon in the Mogollon Mountains, Leitendorf in the Pyramid Mountains, Gold Hill in the

Burro Mountains, Old Hachita in the Little Hatchet Mountains, Copperton in the Zunis, Kelly in the Magdalenas, Cookes in the Cookes Range, Chance City in the Victorio Mountains, and many, many more.

Like the mines that created them, almost all of these mining camps were ephemeral.

Nonetheless, by 1900 minerals worth $110 million had been extracted in New Mexico, about one-third of that gold and silver; coal accounted for $40 million of the total. Totals since 1900 have been dramatically higher, coinciding with a shift from precious metals to industrial minerals such as zinc, manganese, fluorite, low-grade copper ores, potash, petroleum, molybdenum, and uranium. These, too, however, often are of transient importance, though the effects of mining, including disfigurement of landforms and leaching of toxic chemicals into water sources, can be very long lasting.

At present, mining is at an ebb in New Mexico's mountains. Mining for precious metals belongs more to history than economic development, as does mining for coal. The large surface mines for copper near Silver City and molybdenum near Questa struggle with adverse market conditions. Despite this, however, New Mexico's mountains continue to hold considerable mineral wealth, and mining will remain an important part of the history of New Mexico's mountains.

Southern Rocky Mountains

he Rocky Mountains, North America's mightiest mountain chain, cleaver of the continent, has its southern end in New Mexico, terminating east of Santa Fe, north of I-25. Here the great wrinkle in Earth's crust is manifest as the Sangre de Cristo Mountains and includes the Culebra, Latir, Taos, Picuris, Cimarron, and Santa Fe subranges. It is the largest and highest mountain complex in New Mexico. On the list of New Mexico's tallest mountains, *all* of the first seventy-one are in the Rocky Mountains, including Wheeler Peak, 13,161 feet, the state's highest mountain.

But the Southern Rocky Mountain Province in New Mexico includes more than the Rockies. Bifurcated by the deep structural depressions along the Rio Grande Rift, the province is bounded by the Sangre de Cristo Mountains on the east and by the Jemez, Tusas, and Nacimiento Mountains on the west. The latter two groups have Precambrian crystalline cores overlain by later sedimentary formations; the Jemez Mountains resulted from relatively recent (late Cenozoic) volcanism. The Rio Grande Rift also is the channel through which the Rio Grande flows, and this subsiding sliver of Earth's crust, between parallel faults corresponding to volcanism and uplift, runs the length of New Mexico.

Ecologically, latitude and especially high elevations make this province the state's coldest and wettest. Only here is found the alpine life zone; only here grow bristlecone pines. The animals here also are more typical of Colorado than of New Mexico. The Sangre de Cristo and nearby Jemez and Tusas Mountains are the southernmost limit of martens, pikas, and yellow-bellied marmots, among other species.

The state's largest wilderness array is in this province: the Pecos,

Southern Rocky Mountains in New Mexico

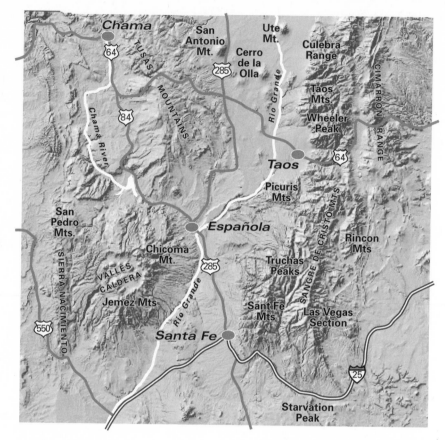

Wheeler Peak, Latir Peak, Cruces Basin, Chama River Canyon, San Pedro Parks, and Dome Wildernesses, as well as the Columbine-Hondo and other Wilderness Study Areas, the Valles Caldera National Preserve, and the Bandelier Backcountry.

Because of their extent and available moisture, both from winter snow-pack and summer monsoon rains, the Southern Rocky Mountains (in southern Colorado as well as northern New Mexico) give rise to almost all of New Mexico's major rivers, including the Rio Grande, San Juan, Animas, La Plata, Pecos, Red, Santa Fe, Jemez, Gallina, Rio los Pinos, Mora, Sapello, northern and east Rio Puercos, and Canadian. This is the only region in New Mexico where a blue line on a map might actually mean flowing water!

Sangre de Cristo Mountains

Location: Northern end in Colorado, south of Salida, extending into
    north-central New Mexico, southern end east of Santa Fe near Pecos.
Physiographic province: Southern Rocky Mountains.
Elevation range: 7,000–13,000 feet.
High point: Wheeler Peak, 13,161 feet; in Colorado, Crestone Peak, 14,294 feet.
Other major peaks: See subranges following.
Ecosystems: Piñon-juniper, ponderosa pine, Douglas fir, white fir, blue
    spruce, Engelmann spruce, aspen, subalpine fir, alpine grasses
    and sedges.
Dimensions: 220 miles total length, including Colorado. See New
    Mexico subranges below.
Counties: Taos, Santa Fe, Colfax, Mora, San Miguel.
Administration: Carson and Santa Fe National Forests, BLM—Santa Fe
    and Taos Field Offices, land grant, tribal, state, private.
Indian lands: Taos, Picuris, Nambe, and Tesuque Pueblos.
Wilderness: Pecos Wilderness, 223,333 acres; Latir Peak Wilderness,
    20,506 acres; Wheeler Peak Wilderness, 19,150 acres; Columbine-
    Hondo WSA, 30,500 acres.
Getting there: See individual subranges.

The Sangre de Cristo Mountains are New Mexico's greatest range, and the southern extremity of North America's greatest mountain chain—the Rocky Mountains. On the list of New Mexico mountains, the tallest forty-three all are in the Sangre de Cristos, including *all* of the state's peaks twelve thousand feet or higher. At one hundred miles, the range is the state's longest, not even counting the portion that is in Colorado. Indeed, even in Colorado, with numerous major mountain ranges, the Sangre de Cristos are among the most significant, with 8 of the state's 53 Fourteeners (lists vary), and the 226,455-acre Sangre de Cristo Wilderness, as large as the Pecos Wilderness.

Physiographically, the Sangre de Cristo Mountains are part of an eastern belt of the Southern Rocky Mountains of Colorado and northern New Mexico. The arc-like range is bordered on the west by the Rio Grande Rift and on the east by the Raton Basin. The Sangre de Cristo Mountains are more complex in New Mexico than in Colorado, where the range is a long, narrow crest. Here, the crest is less well-defined. North of Jicarilla Peak the range consists of a broad, nearly flat ridge deeply incised by glacial valleys. South of Jicarilla Peak, the range splits into two complex branches, separated by the Pecos River. The eastern mountains, exemplified by Elk Mountain, are rather gentle, with no sharp peaks. The western summits, such as the Truchas Peaks, are steeper, and much more rocky and rugged.

Like the rest of the Rocky Mountains, the Sangre de Cristo Mountains were reincarnated during the Laramide Orogeny (mountain building) about 70 million years ago, rejuvenated in the Miocene (23 million years ago), and are still growing vigorously today. The mountains northeast of Taos abruptly rise from the plains in the bold exuberance of adolescence. Yet while the mountains are young, the rocks comprising them typically are not, especially in the range's western sections. Precambrian rocks dating from 1.6 billion years ago are visible in road cuts and atop the range's highest peaks. As William deBuys has noted of these rocks, "Their most salient common characteristic is their hardness, and affects every aspect of the land that they form. Hard rock means slow soil formation and steep topography" (*Enchantment and Exploitation: the Life and Hard Times of a New Mexico Mountain Range*). Not all the rocks are ancient, however; near Red River, for example, the ancient igneous and metamorphic rocks are overlain by Neogene volcanics a mere 25 million years old, and lavas on the eastern slopes are as young as 2 million years old.

About fifteen thousand years ago, the Sangre de Cristo Mountains assumed their present configuration, when Pleistocene glaciers began retreating, leaving behind characteristic U-shaped valleys, linear moraines, and bowl-like cirques, quarried by the grinding and churning of high-mountain glaciers. No other New Mexico mountains owe so much of their topography to glaciation.

Ecologically, the Sangre de Cristo Mountains also are the southernmost expression of the Rocky Mountains. Bristlecone pines reach their southern limit here. So do animals such as pine martens, pikas, yellow-bellied marmots, heather voles, and snowshoe hares.

The earliest humans in the mountains doubtless were nomadic hunter-gatherers, not unlike the Utes the Spanish found here in the seventeenth century. The Sangre de Cristos also were within the traditional territory of the Jicarilla Apaches. They believed the range was a spine running down the center of their homeland, with the heart somewhere near Taos Pueblo. The mountains also have been within the traditional territories of numerous Puebloan groups, several of whose pueblos are still inhabited in the mountains' foothills; see Taos Mountains and Picuris Mountains. Individual mountains and other sites figure in the mythologies of all the pueblos, who have held ceremonies and maintained shrines in the mountains (see *Blue Lake* below).

In 1719, the Spanish explorer Antonio Valverde y Cosio visited the mountains and supposedly named them *Sangre de Cristo* after seeing the mountains turn pinkish-red at dawn and dusk. Nonetheless, perhaps no New Mexico name has been more subject to apocryphal folk-etymologies. The most popular is that of the Spanish priest who, mortally wounded during the Pueblo Revolt of 1680 and asking for a sign from God, exclaimed "Sangre de Cristo!" when with his dying breath he beheld the mountains red at sunset. Scholar T. M. Pearce believed the name dated from the early nineteenth century and reflected the popularity of the Penitente religious fraternity, which emphasized the passion of Christ. As *The Place Names of New Mexico* explains,

> Before this, early Spanish documents referred to the mountains
> as *La Sierra Nevada*, "the snowy range"; *La Sierra Madre*, "the
> mother mountains"; or simply as *La Sierra*. Early English speaking
> trappers called the range *The Snowies*. Fray Francisco Atanasio
> Domínguez, describing Santa Fe in 1776, said the *Sierra Madre*
> lay to the east and was abundant in firewood and timber. The
> US Board on Geographic Names in 1965 chose to designate
> the group the Sangre de Cristo Mountains and not the *Sangre
> de Cristo Range*.

Late spring in the Truchas Peaks, behind the town of Trampas

Spanish settlement of the Sangre de Cristos was sparse, despite numer-ous land grants here. When José de Urrutia reported on his 1766–68 trip to Santa Fe (he produced the first map of Santa Fe) he wrote: "To the east of the Villa [Santa Fe], about one league distant, there is a chain of very high moun-tains which extends from south to north so far that its limits are unknown."

Like the Puebloans, Hispanics established their villages primarily around the mountains' periphery, but they used the mountains extensively for fuel wood, hunting, and most importantly, grazing of sheep and cattle, traditional uses still practiced. Hikers in the Pecos Wilderness must accept sharing the meadows with cows, and William deBuys, writing about the Sangre de Cristos, cites as evidence of grazing impact the wildflower-filled cirque on Santa Barbara called *Rincon Bonito* ("pretty corner"); "It owes its floral display to tens of thousands of sheep eradicating the grass, leaving only inedible forbs." (*Enchantment and Exploitation: the Life and Hard Times of a New Mexico Mountain Range*).

Little mining occurred in the Sangre de Cristo Mountains during the Spanish and Mexican periods, but that changed significantly when Americans began arriving after 1850. Even figures such as Ceran St. Vrain and Kit Carson were swept up in the mining enthusiasm; on April 1, 1865, Carson and three others filed claims on the Kit Carson Lode, near Arroyo Hondo. Red River, Twining, Elizabethtown, Tererro, and numerous other camps were born during this time.

But by the early twentieth century mining and the mining camps were declining. Many settlements, such as LaBelle and Elizabethtown, were abandoned completely. Others, such as Red River and Twining, survived long enough to tap a lode far more plentiful and valuable than precious metals: snow for skiers.

Americanization also had a dramatic impact on landownership in the mountains. Former land grants became part of the National Forests, though a few grants have survived more or less intact, including the Sangre de Cristo Land Grant near the Colorado border. The status of land grants has remained a contentious issue unto the present.

But of all the historic developments in the Sangre de Cristos, the most significant was the establishment of the National Forests. In the waning years of the nineteenth century, Americans finally were awakening to the limits of the nation's natural resources, and a consensus was building that formal protection was necessary to preserve them and head off such consequences as watershed degradation, flooding on unprotected soils, and loss of wildlife habitat and the animals themselves. In 1891, 12 million acres of forest were consumed by forest fires in the Sangre de Cristo Mountains. Then in 1892 President Harrison proclaimed the upper Pecos watershed a timberland reserve for watershed protection (a proclamation not implemented until 1898). The area was set aside and withdrawn from every use including logging, grazing, and mining—and it was closed completely to the public. It was the nation's second National Forest. In 1906, the Carson National Forest was established, bringing most of the Sangre de Cristo Mountains into public ownership. The exclusion of the public was short-lived, but the Santa Fe and Carson National Forests continue their stewardship of most of the Sangre de Cristo Mountains in New Mexico.

Forest protection was followed soon by wildlife restoration, and a central figure in this was Elliott Barker. Raised on a ranch in the southeastern slopes of the Sangre de Cristo Mountains, Barker knew the mountains intimately, and when he joined the Forest Service in 1918 he came in contact with others who shared his concern about the mountains' welfare. Young Barker's supervisor on the Carson National Forest was Aldo Leopold.

In 1931 Barker was appointed state game warden, a position he held for twenty-two years. During that time he changed the face of the New Mexico mountain wilderness and was especially active in the restoration and protection of wildlife. He attempted reintroduction of Rocky Mountain bighorn sheep into the Sangre de Cristo Mountains (attempts succeeded only in 1967, partly due to the removal of domestic sheep, which can spread disease to bighorns). He reintroduced beavers into the Pecos country. He restocked elk in areas where they had been exterminated, though native Merriams elk subspecies is gone forever. He established numerous fisheries and worked for watershed protection. He was instrumental in the public acquisition of key land in Cimarron Canyon and along the Pecos River. He lobbied effectively for better game protection and management laws, and helped organize the National Wildlife Federation. He even was instrumental in Smokey Bear becoming the national fire prevention symbol.

Barker, who died in 1981 at 101, in his lifetime saw enormous changes in the Sangre de Cristo Mountains. At age twenty-five, when Barker was an assistant forest ranger, the Pecos High Country was estimated to receive only about three hundred recreationists a year; today they often exceed that number in a single day.

Culebra Range

Location: Continues into New Mexico from Colorado and extends to
  Costilla Creek, east of NM 3.
Physiographic province: Southern Rocky Mountains.
Elevation range: 8,400–12,600 feet.
High point: In Colorado, Culebra Peak, 14,047 feet; in New Mexico,
  unnamed summit 1.2 miles south of Big Costilla Peak, 12,931 feet.
Other major peaks: State Line Peak, 12,867 feet; Big Costilla Peak, 12,739
  feet; Little Costilla Peak, 12,584 feet.
Dimensions: 9 x 11 miles in New Mexico.
Ecosystems: Ponderosa pine, Douglas fir, white fir, blue spruce,
  Engelmann spruce, subalpine fir, and alpine grasses and sedges.
Counties: Taos.
Administration: Sangre de Cristo Land Grant; permit is required from

the Rio Costilla Co-op Livestock Association: P.O. Box 111, Costilla, NM 87524, (505) 586-0542; Carson National Forest— Questa Ranger District.

Getting there: From NM 3 at Costilla, NM 196 heads southeast toward the mountains through the village of Amalia and continues to follow Costilla Creek along the range's southern border.

One of the major ranges of the Southern Rocky Mountains, the Culebra Range's northern end is in Colorado, south of La Veta Pass, and is a high ridge along which are more than ten summits exceeding 13,000 feet, including 14,047-foot Culebra Peak. In New Mexico, the ridge is slightly lower, but State Line Peak is the state's tenth highest summit, and a hiker along the alpine ridge flanked by glacier-carved valleys and cirques would not notice leaving Colorado and entering New Mexico. (Actually, this hypothetical hiker would be an anomaly, as the range is privately owned in both states, though permission may be granted upon request.)

The Spanish name *Culebra* means "snake," believed to have been inspired not by the mountains but by sinuous Culebra Creek in Colorado, labeled the *Rio de la Culebra* on Lieutenant Zebulon Pike's map of 1810.

Cimarron Range

Location: East of the Moreno Valley, from the Valle Vidal south to Agua
    Fria Creek.
Physiographic province: Southern Rocky Mountains.
Elevation range: 8,800–12,000 feet.
High point: Baldy Mountain, 12,441 feet.
Other major peaks: Touch-Me-Not Mountain, 12,065 feet; Waite Phillips
    Mountain/Clear Creek Mountain, 11,741 feet; Tolby Peak, 11,527 feet;
    Comanche Peak, 11,299 feet.
Dimensions: 15 x 40 miles.
Ecosystems: Ponderosa pine, Douglas fir, white fir, blue spruce,
    Engelmann spruce.
Counties: Colfax.
Administration: State (Colin Neblett and Elliot Barker Wildlife Areas and
    Cimarron Canyon State Park), Philmont Scout Ranch, Carson
    National Forest—Questa Ranger District, private.

Getting there: Access to state lands along the Cimarron River is via US 64 through Cimarron Canyon. Also from US 64, east of the village of Cimarron, Forest Road 1950 heads northwest into the Valle Vidal and northern portions of the range.

The Cimarron Range is the easternmost range in the Sangre de Cristo Mountains. As such, the mountains were the interface between the cultures and resources of the Plains and those of the mountains. The Utes, primarily a mountain people, lived here, as did the Jicarilla Apaches. But the mountains also were along trading and hunting routes used by Rio Grande Puebloans and, later, Hispanics, as well as Plains Indians such as the Kiowas and Comanches. For similar reasons, the Cimarron Range attracted American traders and trappers when they entered the area in the early 1800s. It was at Rayado in the range's eastern foothills, south of Cimarron, that Kit Carson made his home.

The Mountain Man era was colorful but brief, its main legacy the present landownership pattern. In the final days of Mexican administration, New Mexican governor Manuel Armijo allotted huge land grants, often to individuals. In 1841 he gave such a grant to Carlos Beaubien, a French trapper, and Guadalupe Miranda, of Taos. Through marriage and purchase, the entire grant became owned by Kit Carson's friend, Lucien Bonaparte Maxwell, who almost overnight went from mountain man to land baron, owner of the largest private tract in the Western Hemisphere, 1,714,765 acres. Eventually Maxwell sold his vast holdings to foreign investors, but before that numerous other frontiersmen, little concerned with the niceties of ownership or title, settled in the area. The foreign investors demanded that the settlers leave, or at least pay for their properties. The settlers, not inclined or able to do so, resisted, sparking the armed conflict known as the Colfax County War. Despite this turmoil, some units of the Maxwell lands have remained intact, such as the three hundred fifty thousand-acre Vermejo tract, currently owned by media magnate Ted Turner, and the one hundred thousand-acre Valle Vidal. It was during Maxwell's tenure that the U.S. government established the Ute Indian Agency at Ute Park to serve Utes and Jicarilla Apaches, giving Maxwell a lucrative monopoly on supplying the agency.

At about the same time, in 1866 a party of Utes arrived at Fort Union and offered to pay for supplies with rich copper ore. Two men, W. Kroenig and W. Moore, saw the ore and paid the Utes to lead them to its origin, which was near the summit of Baldy Mountain. There they opened the Mystic Mine. Kroenig and Moore hired three men to do the assessment work, and one of them, named Kelly, found gold when panning gravels in Willow Creek coming off Baldy Mountain. Soon the other two men also panned gold—and the copper was forgotten. In the ensuing boom the mining camps of Baldy and Elizabethtown, named for a miner's daughter, were born. The district was briefly rich in the late 1860s and early 1870s, producing an estimated $2 million worth of gold in its placers, and perhaps even more in lode deposits—a fabulous amount for New Mexico at the time. But eventually the ores played out, the miners moved on, and the camps were abandoned. Elizabethtown became a ghost town; Baldy, now within Philmont Scout Ranch, was razed in 1941.

Eagle Nest Lake, an irrigation project, was completed in 1919. Then in 1938 Waite Phillips, an Oklahoma oil tycoon, purchased a third of a million acres of the former Maxwell Land Grant. Phillips, a strong proponent of the Boy Scouts, donated one hundred twenty-seven thousand acres to create Philmont Scout Ranch. The ranch included Kit Carson's home in Rayado, restored in 1949 by the Scouts. Today, thousands of Boy Scouts arrive annually at Philmont, which has become a major institution of American Scouting.

Taos Mountains

Location: Northeast of Taos, extending north almost to Colorado,
  separated from the Culebra Range by Costilla Creek.
Physiographic province: Southern Rocky Mountains.
Elevation range: 7,400–13,000 feet.
High point: Wheeler Peak, 13,161 feet.
Other major peaks: Old Mike Peak, 13,113 feet; Simpson Peak, 12, 976 feet;
  Lake Fork Peak, 12,881 feet; State Line Peak, 12,867 feet; Big Costilla
  Peak, 12,739 feet; Gold Hill, 12,716 feet; Latir Peak, 12,708 feet.
Dimensions: 18 x 42 miles.
Ecosystems: Piñon-juniper, ponderosa pine, Douglas fir, white fir, blue
  spruce, aspen, subalpine fir, Engelmann spruce, alpine grasses
  and sedges.
Counties: Taos, Colfax.
Administration: Carson National Forest—Questa Ranger District.

Lake Fork Peak in June

Indian lands: Taos Pueblo is within the Taos Mountains.

Wilderness: Latir Peak Wilderness, 20,506 acres; Wheeler Peak Wilderness, 19,150 acres; Columbine-Hondo WSA, 30,500 acres.

Getting there: From Taos, NM 522 and then 150 lead to Taos Ski Valley, near the western base of Wheeler Peak. From Questa, NM 38 heads east to Red River and eastern approaches to Wheeler Peak. The Latir Peak area also is reached from Questa, via Forest Road 134. On the south and east, US 64 connects with NM 38 at Eagle Nest.

I recall the first time I saw the Taos Mountains. From across the Taos Plain I beheld massive, high mountains rising tier upon tier above the forest of spruce-fir, to far above timberline, to 13,000-foot summits still draped with snow. I all but gasped. Here was the Rocky Mountain High Country I'd loved so deeply growing up in Colorado. I was home.

The Taos Mountains are the not-by-much highest mountains along the one hundred-mile, north-south axis of New Mexico's Sangre de Cristo Mountains. Wheeler Peak is the highest summit in the group, and the Taos

Wheeler Peak and Williams Lake in June

Mountains include many other significant peaks, including Old Mike, at 13,113 feet the second highest peak in New Mexico. Hikers ascending Wheeler Peak begin in a forest of Engelmann spruce, blue spruce, Douglas fir, and perhaps bristlecone pines. They climb to a zone where the trees become smaller, the zone of subalpine fir, and higher still they encounter dwarf, wind-pruned krumholz pines. At timberline, they walk across wildflower-dappled meadows and emerald marshes, the true alpine zone. Hikers will likely hear pikas barking as the diminutive furry creatures (think guinea pig) scurry among the granite talus. Hikers should also look for yellow-bellied marmots sunning themselves on flat boulders. Still higher they'll be on tundra, the tough turf of low plants adapted to the intense sun, short growing season, deep cold, and desiccating wind of high elevations.

Geologically, the Taos Mountains were born with the general Southern Rockies-Sangre de Cristo uplift, described earlier. This brought to the surface some of New Mexico's oldest rocks. In road cuts along NM 38 between Questa and Red River is foliated quartz monzonite, 1.5 billion years old. Rocks in the Gold Hill area have been estimated at 1.8 billion years old. The uplift revealed other Precambrian metamorphic rocks as well—gneisses, schists, quartzite, amphibolite, and others. Yet nearby, in the Red River Valley, are volcanic rocks from the Paleogene-Neogene transition, a mere 25 million years old.

Much more recently, a mere fifteen thousand years ago, Pleistocene glaciers quarried glacial cirques and gouged characteristic U-shaped valleys. Advancing and retreating, the glaciers left moraines, such as those just north and west of Williams Lake, at the western base of Wheeler Peak.

The name Williams Lake likely recalls Old Bill Williams, quintessential mountain man, but he was a relative latecomer in the mountains. From time out of mind prehistoric Indian groups gathered in the shadow of the Taos Mountains, at the site of present Taos and Taos Pueblo, to trade, buy supplies, maintain social contacts, and more. After Europeans arrived, the tradition continued, and by the time Kit Carson was living in Taos, the annual rendezvous was famous throughout the West as a magnet for Native tribes, mountain men, fortune-seekers, outlaws, and above all, traders. Here captives and slaves might be traded, and possibly reunited with their families. Imagine the Taos Rendezvous as Carson would have seen it, the sprawling meadows at the mountains' base crowded with teepees and camps, remudas of horses, and people chattering, shouting, drinking, boasting, haggling in a score of languages, while the mountains behind stood aloof and silent.

Despite early Spanish familiarity with the Taos Mountains and early mission settlements associated with Taos Pueblo, raids by nomadic Indians retarded Hispanic settlement here. Though the Taos Indians are the group most intimately associated with the Taos Mountains, sharing their name with them, the mountains also were within the hunting and raiding territories of Utes, Navajos, and Jicarilla Apaches, as well as Comanches and other Plains Indians.

Mining came late to the Taos Mountains. Prospectors had probed the area, beginning around 1869, but not until the 1890s did significant mining begin, spawning numerous mining camps: Twining, Red River, Midnight, LaBelle, Amizette, and others. All proved ephemeral, except Red River and Twining (now Taos Ski Valley). Mining continues in the Taos Mountains at the Molycorp molybdenum mine near Questa.

Now tourism is the main industry. Major ski areas are at Taos Ski Valley and in Red River. Fishing and hiking are popular in the summer, especially in the High Country of the Latir Peaks Wilderness, the Columbine-Hondo

Foggy trail to Wheeler Peak

Wheeler Peak

People viewing the mountains from the west, from near Taos, gaze upon a high, snow-capped peak and assume it is Wheeler Peak. But more likely it is 12,305-foot Pueblo Peak; Wheeler isn't visible from Taos. In fact, the only good view of Wheeler is from the east and the Moreno Valley. Despite its height, and unlike mountains such as Mount Taylor, and Cookes Peak, Wheeler doesn't present a distinctive profile anyone would recognize. The Bull-of-the-Woods Trail route to Wheeler is scenic and satisfying but not particularly challenging. The Williams Lake route is shorter but also steeper.

For decades Wheeler Peak wasn't even accorded the distinction of being New Mexico's highest summit. Certainly Major George M. Wheeler, the army surveyor for whom the peak was named, didn't regard it as the highest. Everyone knew that Truchas Peak (South Truchas) was the highest. Yet in 1948, the Santa Fe mountaineer and photographer, Harold D. Walter, borrowed some surveying equipment, shot Wheeler, and showed it edging out the Truchas Peaks. Subsequent surveys proved him right. *New Mexico's Wilderness Areas* finishes the story: "After Walter's death in 1958, a previously unnamed prominence just north of Wheeler's summit was officially named Mount Walter, a name Walter himself had begun using." A plaque on the summit commemorates Walter, "who loved these mountains."

WSA, and the Wheeler Peak Wilderness, where reaching the summit of New Mexico's highest mountain is an obvious objective. The most direct route to the top, useful if one wants to be up and down early to avoid afternoon monsoon thunderstorms, is from Williams Lake, but the route is very steep. The 7.2-mile Bull-of-the-Woods route is longer but more gradual. And the 10-mile route from Red River is longer still but is an outstanding backpack trip.

Latir Peak Wilderness and Columbine-Hondo WSA

Overshadowed by the Wheeler Peak Wilderness just to the south, these two wild areas often are overlooked by hikers. The Columbine-Hondo Wilderness Study Area includes 30,500 acres of conifer forest and high-elevation tundra, culminating at Gold Hill, 12,716 feet. The WSA lies north of the Rio Hondo/NM 150 and south of the Red River/NM 38. During the mining boom before and after 1900, miners and prospectors traveled often here,

Venado Peak in the Latir Peaks Wilderness

and the Columbine-Twining National Recreation Trail, linking the Forest Service Campgrounds, Columbine and Twining, follows an old route between the Red River and the Rio Hondo.

The Latir Peak Wilderness, 20,506 acres, is centered on a cluster of 12,000-foot summits: Venado Peak, 12,734 feet; Latir Peak, 12,708 feet; Virsylvia Peak, 12,594 feet; and Cabresto Peak, 12,448 feet. Most hikers approach the mountains from the south and Cabresto Lake, reached from the village of Questa by taking NM 38 east just a short distance, then branching northeast on NM 522, which is paved until it reaches the Carson National Forest boundary, where it becomes good dirt Forest Road 134. An outstanding, long day-hike or backpack is the fourteen-mile loop that begins at Cabresto Lake, goes along the Lake Fork of Cabresto Creek to Heart Lake, then climbs to Latir Mesa and the above-timberline peaks. From the Latir Mesa summit, the route descends via Trail 85. At the junction with Trail 88, Trail 85 descends along Bull Creek to rejoin the Lake Fork Trail and return to Cabresto Lake, a total of fourteen miles. Also from NM 522,

Battle for Blue Lake

It was in the Taos Mountains that Indians fought one of their most important battles, the battle for Blue Lake, but unlike earlier Indian conflicts, this was not fought by warriors with weapons but rather by lawyers and lobbyists—and the Indians eventually won.

Located at the southeast base of Wheeler Peak, Blue Lake is sacred to the people of Taos Pueblo, who conduct ceremonies there not to be witnessed by outsiders. Yet Blue Lake was included in the Carson National Forest when it was established, and Taos Indians in the early twentieth century watched with increasing dismay as more and more fishermen, hikers, picnickers, and other recreationists began visiting the lake, littering and degrading the site. The pueblo petitioned the United States government to get Blue Lake back. In courtrooms and in Congress, with lawyers and lobbyists, and with major non-Indian allies in New Mexico, the Indians fought for sixty years before finally, in 1970, Blue Lake and forty-eight thousand surrounding acres were ceded to the pueblo. Author William deBuys captured the moment's significance when he wrote: "The celebration that followed was the celebration of a people who had survived a cultural apocalypse" (*Enchantment and Exploitation: the Life and Hard Times of a New Mexico Mountain Range*). In 1997, the nearby area known as the Bottleneck, between Old Mike and Simpson Peaks, also was transferred back to the pueblo.

Trail 88 enters the wilderness from the northwest by following the Rio Primero (First Creek). After about 4.5 miles, Trail 88 joins Trail 85, passing between 11,948-foot Pinabete Peak and 12,448-foot Cabresto Peak en route.

Fernando Mountains

Location: East of Taos, south of the Rio Fernando de Taos, north of the Rio Chiquito.

Physiographic province: Southern Rocky Mountains.

Elevation range: 7,200–10,000 feet.

High point: Sierra de Don Fernando, 10,365 feet.

Other major peaks: Cerrito Colorado, 10,246 feet.

Dimensions: 4 x 11 miles.

Ecosystems: Ponderosa pine, Douglas fir, white fir, and blue spruce.

Counties: Taos.

Administration: Carson National Forest—Camino Real Ranger District.

Getting there: From near the mouth of the canyon of the Rio Fernando de Taos, Forest Road 164 heads east-southeast along the Fernando Mountains' crest, intersecting several Forest Roads and trails.

Like the Picuris Mountains farther south, the Fernando Mountains are comprised of the ridges and valleys branching from a sinuous ridge trending east-west. They take their name from the river along their northern border, named in turn for a settler, Don Fernando de Chavez, prominent in the area before the Pueblo Revolt of 1680.

Picuris Mountains

Location: North of Picuris Pueblo, east of Pilar.

Physiographic province: Southern Rocky Mountains.

Elevation range: 6,000–10,500 feet.

High point: Picuris Peak, 10,801 feet.

Other major peaks: Cerro de las Marqueñas, 7,641 feet.

Dimensions: 10 x 15 miles.

Ecosystems: Piñon-juniper, ponderosa pine, and spruce-fir.

Counties: Taos.

Administration: Carson National Forest—Camino Real Ranger District; BLM—Taos Office; Picuris Pueblo; Cristoval de la Serna and Gijosa land grants.

Indian lands: Picuris Pueblo lands include the mountains' southern and western sections.

Getting there: From Vadito and NM 75, Forest Road 469 heads north toward Picuris Peak.

Bounded on the west and north by the Rio Grande, on the south by the Rio Pueblo, and on the east by NM 518, the Picuris Mountains have been described by geologists as a "wedge-shaped prong of Precambrian crystalline rock that extended westward from the Sangre de Cristo Mountains to the Rio Grande Rift." The mountains are characterized by high, steep

The Harding Pegmatite Mine

The Picuris Mountains are famous among geologists and rockhounds for two things: "fairy crosses," twinned interpenetrating crystals of staurolite, and the Harding Pegmatite Mine. In the late Precambrian, about 800–900 million years ago, granitic dikes and veins intruded older igneous rocks. Known as pegmatites, these intrusions cooled and crystallized slowly beneath the surface, allowing large crystals to form. The Harding pegmatites happened to be, as one geologist put it, "a unique occurrence of unusual minerals."

The mine was opened about 1910 and presumably named for President Warren G. Harding. Prospectors for precious metals were disappointed by not finding significant gold—the early nineteen hundreds' gold-mining promotion at Glen-Woody near the Rio Grande was just that, a promotion—but the dikes were especially rich in beryl, along with lithium and tantalum-columbium minerals. Between 1950 and 1955, 752 tons of beryl were mined, more than twenty percent of total U.S. production then.

More enduringly, however, the pegmatites yielded mineral specimens of exceptional variety and beauty, including apatite, almandine garnet, and spodumene. Mining has ceased, and in 1974 the mine's owner, Dr. Arthur Montgomery, professor emeritus of geology at Lafayette College in Pennsylvania, donated the mine to the UNM Department of Earth and Planetary Sciences, which administers it. A permission-release form from the department is required before entering.

ridges and deep canyons, creating a forested backdrop for the Tiwa pueblo that gave the mountains their name. Elevations rise gradually eastward to culminate at Picuris Peak, recognized by its lookout tower.

Rincon Mountains

Location: North of Mora, between NM 121 and 434, between the
　　valleys of the Mora River and Guadalupita Canyon.
Physiographic province: Southern Rocky Mountains.
Elevation range: 7,600–9,400 feet.
High point: Unnamed summit, 9,570 feet.

Other major peaks: Los Hornos, 9,417 feet; Comanche Peak, 8,980 feet.
Dimensions: 6 x 15 miles.
Ecosystems: Ponderosa pine, Douglas fir, white fir, and blue spruce.
Counties: Mora.
Administration: Private.

This is a small subrange, likely named for its location in a rincón, or "corner," between two valleys.

Santa Fe Mountains

Location: Extend in the north from the Rio Pueblo and Embudo Creek, south of NM 75, south of Peñasco, to Glorieta Pass in the south.
Physiographic province: Southern Rocky Mountains.
Elevation range: 8,000–13,000 feet.
High point: Truchas Peak, 13,102.
Other major peaks: Middle Truchas, 13,066 feet; North Truchas, 13,024 feet; Chimayosos Peak, 12,841; Jicarita Peak, 12,836 feet; Santa Fe Baldy, 12,662 feet; East Pecos Baldy, 12,528 feet; Santa Barbara, 12,515 feet; Pecos Baldy, 12,500 feet; Lake Peak, 12,409 feet; Penitente, 12,249 feet; Sierra Mosca, 11,801 feet; Elk Mountain, 11,661 feet; Hermit Peak, 10,260 feet; Glorieta Baldy, 10,199 feet.
Ecosystems: Piñon-juniper, ponderosa pine, Douglas fir, white fir, blue spruce, Engelmann spruce, aspen, subalpine fir, alpine grasses and sedges.
Counties: Rio Arriba, Mora, Santa Fe, San Miguel.
Administration: Santa Fe National Forest—Española, Pecos, and Las Vegas Ranger Districts; Carson National Forest—Camino Real Ranger District, Peñasco.
Indian lands: Picuris, Nambe, Pojoaque, and Tesuque Pueblos include portions of the Santa Fe Mountains on the north and west. Pecos Pueblo, now abandoned, was in the foothills on the south.
Wilderness: Pecos Wilderness, 223,333 acres.
Getting there: From the south, the best access is from NM 63 between Pecos and Cowles. From the east, NM 65 from Las Vegas heads toward the Hermit Peak area. From the west, NM 475 from Santa Fe leads to the Santa Fe Ski Basin and the Winsor Trail. From the north, points

On the ridge line between the Pecos Baldy Peaks

of access include Trampas Creek via El Valle, the Santa Barbara
Campground, and the Serpent Lake Trail via Forest Road
161 southeast of Angostura.

With the Santa Fe Mountains, the great chain of the Rocky Mountains
comes to its southern end. If you simply eyeball a topographic map, you
might conclude the specific southernmost point is about 1.5 miles west of
Bernal, near I-25 between Santa Fe and Las Vegas. From here, the Santa Fe
Mountains extend approximately sixty miles northward, to end at the Rio
Pueblo and Embudo Creek. Within this ocean of forest are archipelagos of
peaks. Here are the Truchas Peaks, remote and difficult, as well as the
"Baldies"—Santa Fe, Glorieta, East Pecos, and Pecos. Most high mountains
here, such as Lake Peak and Penitente Peak, for centuries have been held
sacred by the many Native American groups who live in and around the
Santa Fe Mountains. Hermit Peak and Elk Mountain reflect the long presence of Hispanics. And here also is the Pecos Wilderness, at 223,333 acres
New Mexico's second largest wilderness. It was in the Santa Fe Mountains
that Elliott Barker (1886–1987), New Mexico's foremost conservationist,
grew up and learned to cherish the values of wild nature. He once estimated he'd ridden one hundred twenty thousand wilderness miles; he took
his last trail ride in the Pecos Wilderness when he was 89. Truly, the Santa
Fe Mountains are a worthy terminus of the Rocky Mountains.

Within the overall range, the highest peaks are clustered into three sub-
regions:

Truchas Region
Location: South of the Rio Pueblo and Embudo Creek, north of
　　Pecos Baldy.
High point: Truchas Peak, 13,102 feet.
Other major peaks: Middle Truchas Peak, 13,066 feet; North Truchas,
　　13,024 feet; Chimayosos Peak, 12,841 feet; East Pecos Baldy, 12,529 feet;
　　Santa Barbara, 12,515 feet; West Pecos Baldy, 12,500 feet; Jicarilla Peak,
　　12,494 feet; Trampas Peak, 12,170 feet.
Ecosystems: Piñon-juniper, ponderosa pine, Douglas fir, white fir, blue
　　spruce, aspen, subalpine fir, Engelmann spruce, alpine grasses
　　and sedges.
Counties: Rio Arriba, Mora, and Santa Fe.

The Truchas Peaks

Administration: Santa Fe National Forest—Española Ranger District,
   Carson National Forest—Camino Real Ranger District in Peñasco.
Getting there: From the village of Truchas, Forest Roads 639 and 400 head
   east. Trail 153 goes toward the peaks along the Rio Quemado. (Note:
   Thefts and vandalism of vehicles have occurred at trailheads here.)

Only the Wheeler Peak Wilderness rivals the Truchas Peaks as New
Mexico's premier high country region. Of the state's ten highest peaks,
four are here—deep in the Pecos Wilderness, beyond the reach of a

simple day hike. The backpack route that goes from Cowles to Pecos Baldy Lake, then over Trail Riders Wall to the Truchas Peaks and the Truchas Lakes is considered among the state's classic hikes. From Truchas Lakes, a trail climbs to the saddle between North Truchas and the cirque that is the head of the North Fork of the Rio Quemado. Follow the narrow, rocky, and somewhat exposed ridge south to the summits of Middle Truchas and then South Truchas.

One attraction of the Truchas Peaks is the resident herd of introduced Rocky Mountain bighorn sheep. I remember once stalking them to get a photo, expecting them to bolt as soon as they sensed my presence. Soon they did look my way—and promptly sidled over to try to share my granola bars! (Do not feed bighorn sheep.) The bighorn sheep reintroduction here, following their extirpation around 1900, has been particularly successful, and the herd has grown to provide animals for restocking in other areas, including Wheeler Peak.

Las Vegas Region

Location: East of the Pecos River, west and south of NM 518.

High point: Unnamed high point near Gascon Point, 11,931 feet.

Other major peaks: Elk Mountain, 11,661 feet; Mount Barker, 11,445 feet;
Spring Mountain, 11,180 feet; Hermit Peak, 10,260 feet; El Cielo
Mountain, 9,830 feet; Barillas Peak, 9,362 feet.

Ecosystems: Piñon-juniper, ponderosa pine, Douglas fir, white fir, blue
spruce, Engelmann spruce, aspen, subalpine fir.

Counties: San Miguel, Mora.

Getting there: From Las Vegas, NM 263 heads east to El Porvenir and
nearby campgrounds and trailheads, including those to Hermit Peak
and Elk Mountain.

The mountains here rise steeply from the east, ascending via valleys and ridges to the high, north-south ridge along which are the region's highest summits. This topography was influenced by reverse faults on the east side of the range that occurred during the Laramide Orogeny, 70 million years ago, at the end of the Mesozoic Era. The mountains here typically are broad and forested, rarely rising above true timberline, yet human activity (sheep and cattle grazing, eating the young trees) and natural factors (fires) have created a patchwork of meadows and forest that is excellent wildlife

Hermit Peak

In 1863, a remarkable man arrived in the Las Vegas area, having walked over the Santa Fe Trail from Council Grove, Kansas. He was Juan Maria de Agostini, a well-educated native of Italy who, out of religious conviction, had renounced his family's wealth and set out to become an anchorite in the New World. He took up residence in a cave 250 feet below the rim of the peak that local Spanish-speaking people had called *El Cerro del Tecolote,* "The Hill of the Owl." There he lived, carving crucifixes and religious emblems, which he traded for food, and as his reputation as a holy man and healer grew so did the number of pilgrims who climbed the mountain seeking his blessing. In 1867 he left his cave to take up residence in another cave in the western foothills of the Organ Mountains. In 1869 he was found there, the victim in a murder that was never solved.

Back in the Las Vegas area, however, he was not forgotten. The mountain where he had lived came to be called *El Solitario* in remembrance of him, and as late as 1965 annual pilgrimages were made to his cave by the local *Sociedad del Eremitano.* Nor have the pilgrimages ceased. Hikers to the cave today find along the route crude crosses carried by penitents and at the cave itself offerings made to small shrines.

habitat. The sheep have been removed because they transmit disease to the native bighorns, but cattle are as common as elk here.

The high country around Elk Mountain is accessible via Forest Roads around Terrero. Mining rehabilitation around Terrero Mine has closed some roads; check for local conditions. From Elk Mountain, Trail 251 (the Skyline Trail) enters the Pecos Wilderness and heads north along the main ridge of the Las Vegas region, passing over 11,180-foot Spring Mountain and continuing north along the ridge approximately fourteen miles before turning west to begin climbing sharper ridges to reach the high peaks of the Truchas region. Eventually this trail loops all the way around to Lake Peak.

Most day-hikers coming from Las Vegas focus on Hermit Peak. The four-mile hike from El Porvenir Campground to the summit is moderately strenuous, but the views amply reward the effort. Also worthwhile is the short but steep hike down to the cave where the hermit lived.

Pecos Baldy and its lake, frozen in the basin behind the skier

Santa Fe Region

Location: From Glorieta Pass north to the Rio Frijoles, east of Cundiyo.

High point: Santa Fe Baldy, 12,622 feet.

Other major peaks: Lake Peak, 12,409 feet; Redonda Peak, 12,357 feet; Penitente Peak, 12,249 feet; Glorieta Baldy, 10,201 feet.

Ecosystems: Piñon-juniper, ponderosa pine, Douglas fir, white fir, blue spruce, Engelmann spruce, aspen, subalpine fir, alpine grasses and sedges.

Counties: Santa Fe.

Administration: Santa Fe National Forest—Española, Pecos, and Las Vegas Ranger Districts.

Getting there: From Santa Fe, NM 475 goes approximately fifteen miles east to Santa Fe Ski Basin, where the Winsor Trail leads to Puerto Nambe and eventually to the heart of the Pecos Wilderness at Cowles.

The east slopes of Santa Fe Baldy in the winter

One of the most popular mountain hikes near Santa Fe is the Borrego Trail, which heads at NM 475 just past Hyde Memorial State Park. And it is on this easy, tranquil hike through silent conifer forest that perhaps more than anywhere else one can appreciate the tremendous changes that have occurred in the mountains east of Santa Fe in the past hundred years. *Borrego* is Spanish for "sheep," and it was over this route that Spanish-speaking shepherds once drove their flocks from high mountain pasturage to markets in the capital. A hundred years ago, the trail would have been loud with the baaing of sheep, the barking of dogs, and the shouts of men and boys.

Today, domestic sheep are absent from the mountains, and any clamor on the trail comes from hikers and mountain-bikers, who rarely speak Spanish.

The mountains of the Santa Fe Region are among northern New Mexico's most popular mountain playgrounds—and for good reasons. They are but an hour's drive from Santa Fe over a scenic paved road ending at the Santa Fe Ski Basin. Along the way motorists pass not only the Borrego trailhead but also the Chamisa trailhead and the route at Aspen Vista leading to the towers atop Tesuque Peak. At the ski basin itself they can connect with the Winsor Trail, which begins near the village of Tesuque and goes along Tesuque Creek over Puerto Nambe, whose lush, grassy meadows are a popular destination for backpackers. These days, the animals most likely to be tethered in the meadows are llamas rather than donkeys. From Puerto Nambe, the Winsor Trail continues and eventually ends at Cowles near the Pecos River.

From the Santa Fe Ski Basin, the Winsor Trail provides access to a very steep side trail leading to Nambe Lake, a tiny, shallow, high-mountain lake nestled in a glacial cirque. Great high country scenery—but no fish.

Also from the Winsor Trail hikers gain access to Santa Fe Baldy, the region's highest summit. The well-marked Skyline Trail 251 leads to a saddle along the southeast ridge, where a tangle of trails follows the ridge to the summit—a long but very satisfying day hike. From Santa Fe Baldy's summit hikers look down upon Lake Katherine, one of New Mexico's largest and most scenic glacial lakes. Unlike shallow Nambe Lake, it has trout.

From Puerto Nambe, Trail 251 on the prominent but gradual ridge to the southeast takes hikers to the culturally significant summits of Lake Peak—the Tewas' sacred mountain of the east—and Penitente Peak, where the Catholic religious fraternity *Los Hermanos Penitente* likely maintained a shrine. These two peaks also can be climbed from Tesuque Peak and the top of the ski lifts, but a narrow, dangerous ridge must be traversed.

Less popular than these routes but still satisfying is the five-mile hike to the summit of Glorieta Baldy, from a trailhead in the village of Glorieta. (Forest Road 375 leads to the lookout and campground at Glorieta Baldy.) From Glorieta Baldy a trail goes along a ridge to Thompson Peak, two miles north.

San Antonio Mountain—
Upper Rio Grande Volcanoes

Location: Flanking the Rio Grande, north of Tres Piedras and Taos,
   south of the Colorado border.

Physiographic province: Southern Rocky Mountains.

Elevation: San Antonio Mountain, 10,908 feet; Ute Mountain, 10,093 feet;
   Cerro de la Olla, 9,475 feet; Cerro del Aire, 9,023 feet; Cerro Montoso,
   8,655 feet.

Relief: San Antonio Mountain, 3,000 feet; Ute Mountain, 2,600 feet;
   Cerro de la Olla, 1,870 feet; Cerro del Aire, 1,250 feet; Cerro
   Montoso, 1,000 feet.

Ecosystems: Great Basin grassland, sagebrush, ponderosa pine,
   Douglas fir, and aspen.

Counties: Rio Arriba, San Antonio Mountain; Taos, Ute Mountain,
   Cerro del Aire, Cerro de la Olla, and Cerro Montoso.

Administration: San Antonio Mountain, Carson National Forest—
   Tres Piedras Ranger District; Cerro de la Olla and Cerro Montoso,
   BLM—Taos Field Office; Ute Mountain and Cerro del Aire, private.

Wilderness: North of San Antonio Mountain is a 7,050-acre BLM
   Wilderness Study Area.

Getting there: From US 285 north of Tres Piedras, Forest Road 78 branches
   west and goes beneath the mountain's south side. It connects with
   other dirt roads around the mountain.

It has been stated that San Antonio Mountain is among the nation's largest
"free-standing" mountains—not attached to any range or group. Actually, its
base diameter is a mere five miles, not the largest even in New Mexico.
Mount Taylor is far broader, and taller. Still, the exaggeration is understand-
able. By any standards, San Antonio Mountain is an imposing feature, solid
and massive. It also is a beautiful mountain, its slopes rising in a graceful sig-
moidal arc to the summit. These qualities, accentuated by the surrounding
uninhabited plains, have made San Antonio Mountain a landmark throughout this region. Tewa Indians living along the Rio Grande to the south know
San Antonio Mountain as "Bear Mountain"; it is the mountain of the north
in their cosmology.

San Antonio Mountain encompasses about nine thousand acres and
rises almost three thousand feet, high and broad enough to capture moisture

San Antonio Mountain

and make its own weather. It ascends from sagebrush plains through scrub oak, into ponderosa pines and eventually Douglas fir and shimmering aspens. The mountain's porous volcanic composition precludes surface water and springs, but wildlife managers have set up water tanks for deer, pronghorn, and especially elk, for whom the forest and grassland patchwork is superb habitat. On the mountains' slopes roam some of northern New Mexico's largest elk herds, descendants of the elk reintroduced to the mountain in 1938 and 1939, themselves descendants of animals imported into New Mexico from Wyoming in 1915. All of New Mexico's native elk had been hunted to extinction by 1900, mostly by market hunters, who also decimated New Mexico's pronghorns; by 1917, all of New Mexico had but thirty-five pronghorn bands totaling only one thousand seven hundred animals. Now, winter pronghorn herds sheltering in the forests on San Antonio Mountain have as many as one hundred animals.

Yet, as impressive as San Antonio Mountain is, it is only the best-known and largest—but not by much—of several hulking volcanic masses along the Rio Grande Rift in northern New Mexico. About 7.5 miles south-southeast of San Antonio Mountain is 9,023-foot Cerro del Aire. About eleven

High on the flanks of San Antonio Mountain

miles southeast is 9,475-foot Cerro de la Olla. (Local English-speakers call them Wind Mountain and Pot Mountain.) Farther east is 8,655-foot Cerro Montoso. And east of the Rio Grande, just south of the Colorado border, is 10,093-foot Ute Mountain. All are extinct volcanoes, relics of the tectonic turmoil associated with the spreading Rio Grande Rift that began about 20 million years ago. What a sight it would have been, if one could have accelerated these geological processes—fiery lava oozing over the Taos Plateau, volcanoes in thunderous eruption, choking the air with smoke and ash, as the plains were stretched and torn along the course of the Rio Grande.

There's little hiking on San Antonio Mountain, except during hunting season. A dirt road, gated and locked, goes four miles from US 285 to the top and the electronic communications towers there. Foot travelers, however, would do better just to hike up the mountain's grassy sides, especially on the southeast. The slopes become gradually steeper as they approach the summit, but views along the way evoke the wildness of the Pleistocene. The top is rolling and forested; the highest point is near the communications towers.

A footnote: Testifying to San Antonio Mountain's power to stimulate people's imaginations is its place in the folklore of UFO sightings (as well as cattle mutilations). On December 13, 2000, according to the UFO Folklore Center, "a witness south of Antonito, CO, called to say he was watching a 'huge bright light,' which he described as being located 'three-quarters up San Antonio Mountain.'" Then in 1995 the Rainbow Tribe of Living Light had one of its gatherings on San Antonio Mountain, where thousands of people gathered to connect with nature—and some of them may also have seen lights on the mountain.

Tusas Mountains

Location: North and south of US 64, west of US 285.

Physiographic province: Southern Rocky Mountains.

Elevation range: 6,500–11,000 feet.

High point: Grouse Mesa, 11,403 feet.

Dimensions: 20 x 50 miles.

Other major peaks: Canjilon Mountain, 10,913 feet; Jawbone Mountain, 10,601 feet; Banco Julian, 10,413 feet; Broke Off Mountain, 10,357 feet; Burned Mountain, 10,189 feet; Tusas Mountain, 10,143 feet.

Ecosystems: Piñon-juniper, ponderosa pine, scrub oak, Douglas fir, white fir, blue spruce, Engelmann spruce, aspen, subalpine fir, and Great Basin grassland.

Counties: Rio Arriba.

Administration: Carson National Forest—Tres Piedras and Canjilon Ranger Districts.

Wilderness: Cruces Basin Wilderness, 18,902 acres.

Getting there: US 64 bisecting the range gives access to Forest Roads leading both north and south. The range's western portion is within the Tierra Amarilla Land Grant, with limited access.

One of New Mexico's largest mountain groups, an extension of Colorado's well-known San Juan Mountains, the Tusas Mountains are a somewhat amorphous mass of upland mesas and summits scarcely known outside the local area. No major divide defines the range; it has no mass of peaks, and no major drainages, though several streams head in the Tusas Mountains: the Rios Brazos, Tusas, San Antonio, Cebolla, Nutrias, Vallecitos, and Tierra

The western escarpment of Broke Off Mountain overlooking
the Valle Grande in the Tusas Mountains

Amarilla, as well as the Chaves River. It's a vast, high, green region of gentle mountains and high mesas, of southern Rocky Mountain forests interspersed with meadows and *valles.*

No unity of administration overlays the mountains. The eastern portion is Carson National Forest, the western—Tierra Amarilla Land Grant. The east-west administrative divisions correspond roughly to the mountains' major geological divisions. On the east, the mountains and formations are volcanic. Looking west from the brow of Broke Off Mountain you see that you're standing on the rim of an ancient crater, overlooking a roughly circular caldera. The valley below, Valle Grande, shares more than its name with the Jemez Mountains basin.

The western Tusas Mountains, however, are Precambrian granite and metamorphic rocks. The Brazos Cliffs on the range's western edge owe their verticality to the rock being extremely hard. As climbers put it, the rock has a "good tooth." The Rio Brazos, descending from the high, wet plateau, has incised a deep, narrow canyon in these tough rocks. In the Brazos Box the

The Brazos Cliffs

The Brazos Cliffs, looking down on a climber

river plunges for 3.75 miles between cliffs one thousand four hundred feet high, before finally bursting from the cliffs near the settlement around Brazos Lodge; access is limited by private property here. Upstream, several tributaries join the Rio Brazos through canyons of their own.

Before 1986, the western mountains, including the Brazos Cliffs and the range's highest point, 11,403-foot Grouse Mesa, were called the Brazos Mountain Range, but in that year the U.S. Board on Geographic Names, acting upon the recommendation of the NM Geographic Names Committee, subsumed the Brazos Mountain Range within the Tusas Mountains, ending confusion between the two names for one topographically similar area, despite geological differences. Other subranges of the Tusas Mountains include the Ortega and La Madera Mountains, both near the village of La Madera.

The Tusas Mountains' diverse geology resulted in relatively little mineralization. Pockets of gold and silver, typically low-grade, were discovered in 1881. Neither large nor rich enough to support significant mining, the ore pockets nonetheless yielded several hundred thousand dollars from placers and small mines. Today treasure-hunters with gold pans still arrive at the Rio Vallecitos and the aptly named Placer Creek, hoping to wash a little color from the streams' gravels.

For several reasons, a vanishingly small number of people hike here. Probably the main reason is that as of this writing (early 2002), no formal hiking trails exist in this portion of the Carson National Forest, though the Tres Piedras Ranger District is constructing an extensive system to be called the Tony Marquez Trail, No. 41. Trailheads will be on US 64 near the head of East Gavilan Canyon, north of US 64 near Hopewell Lake, and near Little Tusas Creek at Rincon Negro. Another reason is that the range's highest points are within the land grant and thus require permission to hike. Furthermore, were you to reach the high point on Grouse Mesa, you'd find not a scenic rocky summit but a broad, grassy mesa, likely the summer home of cows. Nonetheless, 10,357-foot Broke Off Mountain, similarly topped by a grassy plateau, is exceptionally scenic when approached from the west. And despite the lack of trails, the network of Forest Roads and ubiquitous meadows make exploration by foot or mountain bike relatively easy.

Hunting and fishing dominate recreation in the Tusas Mountains. Streams here are small, though they often issue from small, high-country lakes, such as the Canjilon Lakes, Trout Lakes, and Hidden Lake. But trout here are plentiful. So are elk; this is superb habitat (see *San Antonio Mountain*). Indeed, elk season is about the only time people visit the Cruces Basin Wilderness. It is among New Mexico's least-visited wildernesses.

Jemez Mountains

Location: North-central New Mexico, south of NM 96, west of US 84,
northeast of US 550.

Physiographic province: Southern Rocky Mountains.

Elevation range: 6,000–10,000 feet.

High point: Chicoma Mountain, 11,561 feet.

Other major peaks: Redondo Peak, 11,254 feet; Polvadera Peak, 11,232 feet;
Cerro Toledo, 10,925 feet; Cerro de la Garita, 10,621 feet; San Antonio
Mountain, 9,978 feet; Cerro Pedernal, 9,862 feet; Saint Peters Dome,
8,463 feet.

Dimensions: 35 x 50 miles.

Ecosystems: Piñon-juniper, ponderosa pine, Douglas fir, white fir, blue
spruce, Engelmann spruce, aspen, and subalpine fir.

Counties: Rio Arriba, Sandoval, Santa Fe, Los Alamos.

Administration: Santa Fe and Carson National Forests; Valles Caldera
National Preserve; Bandelier National Monument; BLM—Rio Puerco
and Farmington Field Offices; tribal administrations; private.

Indian lands: Jemez, Zia, Santa Ana, San Felipe, Santo Domingo, Cochiti,
San Ildefonso, and Santa Clara Pueblos all include portions of the
Jemez Mountains within their current tribal boundaries.

Wilderness: Bandelier Backcountry, 37,000 acres; Dome Wilderness,
5,200 acres; and Valles Caldera National Preserve, 89 acres.

If any mountain group symbolizes New Mexico, it's the Jemez. Mesas,
peaks, extinct volcanoes, ancient and present Indians, Hispanic villages,
Anglo logging and mining camps, remote wilderness, cattle grazing, hunting and fishing, and active tourist attractions, vegetation ranging from
emerald green riparian to dry conifer forests to desert scrub—and at every
turn the unexpected.

Physiographically, the Jemez Mountains are considered part of the
Southern Rocky Mountains, despite their geologic origin being entirely
different from other mountains in the province. Unlike the Sangre de Cristo
Mountains to the east, the Jemez Mountains were not born with a gradually
uplifted granite core but rather with relatively recent volcanism, beginning
with eruptions about 13 million years ago. Three million years ago, coinciding with the spreading Rio Grande Rift, lava flowed from volcanic vents,
creating the dark basaltic layers that characterize so many landforms here;

Jemez Mountains—Sierra Nacimiento

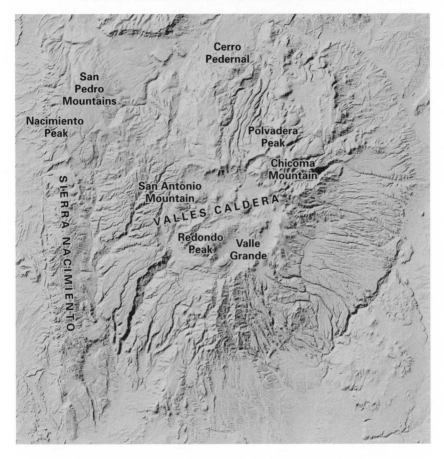

Tent Rocks in the southeastern foothills of the Jemez Mountains

these are especially conspicuous atop the mesas of Bandelier National Monument. The whole geologic drama had a dual climax 1.6 and 1.22 million years ago with two cataclysmic explosions from a huge magma chamber. Vast quantities of ash and debris blanketed the area, settling and coalescing into distinctive, easily eroded tuff deposits. Sometimes boulders

*Redondo Peak overlooking the Valle Grande and the
small dome of Cerro la Jara*

of harder rock were hurled skyward with the ash and incorporated into the
tuff. Later, when the much softer tuff eroded, the boulders protected the
tuff beneath them, resulting in conical tuff formations called "tent rocks."
The best place to see tent rocks and other intriguing tuff formations is
Kasha-Katuwe Tent Rocks National Monument, administered by the BLM,
four miles west of Cochiti Pueblo on Forest Road 466. The short hikes here
are especially appropriate for hikers with children, particularly the route
that threads through a child-sized slot canyon.

Following the volcanic eruptions, the magma chamber, emptied of its
material, collapsed to form the huge, circular basin called the Valles Caldera
at the heart of the Jemez Mountains. The magnificent Valle Grande is part
of the larger Valles Caldera. The high peaks surrounding the grassy *valles*
are domes, such as resurgent Redondo Peak, and the volcanic domes of
Cerro Toledo, Cerro de la Garita, and San Antonio Mountain, pushed
upward by the still-buoyant magma chamber and by remnants of volcanic
activity on the caldera's periphery. The caldera is clearly visible on satellite
images, and the magma body remains beneath the mountains, at a depth

Cross-country skiers approaching Cerro la Jara in the Valle Grande

of about fifteen kilometers—and still rising, albeit slowly. With most of its gasses now gone, any subsequent volcanic activity from the chamber would be as flows rather than as explosions. The area's numerous hot springs are reminders of this magma body's presence.

Today the Valles Caldera is a grassy basin fringed with forest and drained by several small streams, but for much of its history the basin was enclosed and filled with water, a lake. Using drill cores, scientists have studied sediments deposited in the ancient lake to assess climatic fluctuations during its long existence. In the early 1970s, drillers probed the area west of Redondo Peak for its geothermal energy potential.

The Jemez Mountains are a sizable group. The central high mountains on three sides are enclosed within a palisade of basalt-capped mesas. Like fingers from a palm, these mesas radiate outward, separated by very deep, very narrow canyons enclosed within precipitous volcanic cliffs. These finger-mesas are called *potreros*. The Jemez Mountains' northwest flank is guarded by an even more formidable barrier: the linear granite buttress of the Sierra Nacimiento. (Use the letters "IO" to visualize how the two ranges would appear from space.)

Despite their huge catchment area, the Jemez Mountains give rise to relatively few perennial streams, far fewer than the adjacent Sierra Nacimiento. The difference is due to the highly permeable volcanic substrate in the Jemez Mountains. The largest stream is the Jemez River, fed by the East Fork and by San Antonio Creek. More typical are small streams, such as the Rio de los Frijoles. Nonetheless, the mountains are lushly vegetated and support a diverse Southern Rocky Mountain fauna; ermine have been reported in the Valles Caldera, home to one of New Mexico's largest elk herds.

Without doubt the Jemez Mountains, with their rich assemblage of wildlife and edible and medicinal plants, would have been familiar to Paleo-Indians. They left few remains, though evidence of obsidian mining has been found around Cerro del Medio and Rabbit Mountain-Obsidian Ridge and Polvadera Peak. "Weapons grade" obsidian, mostly from the Valles Caldera, has been found at prehistoric sites from Texas to Wyoming.

When people of the Chacoan Culture dispersed into the Colorado Plateau around AD 1150, some groups went east, likely displacing the resident Gallinas Culture. Many settled in the Jemez region; the ruins of their settlements and farmsteads are common throughout the area. By AD 1300, the Puebloan peoples began leaving the mountains and moving into the river valleys, where they live today. But Puebloan peoples were still living on the Pajarito Plateau and on the mesas above Jemez Pueblo when Europeans arrived after 1540. From mesa-top sites such as Tsankawi on the east and Astialakwa on the west, the Indians would have heard rumors of pale, bearded men riding outlandish animals. Now those villages and hundreds of others are but mounds of rubble, though still remembered by modern Puebloans, many of whose ancestors once lived there. The ruins of Tyuonyi in Cañon de los Frijoles are perhaps the most impressive of these ruins, and Bandelier National Monument was created to preserve them and numerous smaller sites in other canyons and on mesa tops. Outside the monument in the Jemez country are countless other sites, large and small. Excavating or disturbing a site, or removing artifacts, is illegal. Most, if not all, of these sites still have meaning for the descendants of the people who created them.

Modern Puebloans also maintain shrines in the Jemez Mountains, and several Jemez peaks are landmarks in their mythologies. Chicoma Mountain, the range's highest summit, has the outlines of a large stone shrine on its top, and most of the peaks themselves have religious significance. Probably the best known of the Jemez Mountains shrines is above Capulin Canyon in Bandelier National Monument, where a low stone wall encircles the barely-recognizable forms of two mountain lions carved in the soft and easily vandalized tuff.

The name of the mountains also recalls the Puebloan presence here. It is evolved from the name the Towa-speaking people at Jemez Pueblo call themselves (just as the Taos Mountains and the Picuris Mountains take their names from the Indians living on their flanks). When Coronado explored along the Jemez River Valley in 1540, he found seven pueblos in the valley and on the mesas above. Now only the pueblo its inhabitants call "Walatowa" remains. But while the settlements have contracted, the name has spread to encompass the mountains, river, springs, falls, and other features.

Early Hispanic settlement of the Jemez Mountains was sparse. Missions had been established at many of the pueblos, but Hispanic settlements beyond the pueblos were limited to a few remote villages—Peña Blanca, Abiquiu, Nacimiento (Cuba), San Ysidro, and others—mostly on the mountains' periphery. Nonetheless, Spanish-speaking people knew these mountains well; almost all the natural features here are labeled with their Spanish names: Pajarito, Frijoles, Alamo, San Pedro, Valle Grande, Gallina, Coyote, among scores more.

The very heart of the Jemez, the Valles Caldera, for most of its history has been known as the Baca Location, or more properly, Baca Location No. 1. When the Mexican government in 1835 and 1841 bestowed two overlapping land grants near Las Vegas, the U.S. government brokered a resolution of the conflict by allowing the second grantor, Juan María Cabeza de Baca, to choose four parcels elsewhere in New Mexico. His first choice was this 155-square-mile tract. In 2000, after decades of negotiations and effort, the Valles Caldera passed into public ownership and now is managed as Valles Caldera National Preserve.

When English speakers arrived in the Jemez Mountains, they were interested primarily in mining and logging. In 1889, gold and silver were discovered in the canyons about twelve miles northwest of Cochiti Pueblo. Mining commenced in 1894, and by 1900 three thousand people were crowded into Bland Canyon, barely sixty feet wide. The boomtown of Bland had four sawmills, two banks, dozens of saloons, a newspaper, a church, an opera house, and even a stock exchange. But, predictably, the ore deposits played out, and Bland, privately owned, is now but a ghost town, as is the neighboring mining camp of Albemarle. Of the former mines, many now are homes to bats. Bars placed across their entrances have kept out humans, but not the bats.

Logging lasted somewhat longer. Like many other New Mexico mountains, the Jemez Mountains were logged heavily in the early twentieth century, and while the trees have long since grown back, roads and railroad grades constructed for logging are still visible. In the western Jemez Mountains, the White Pine Lumber Company in 1922 purchased timber rights to the old Cañon de San Diego Grant, then in private ownership. By the end of 1924 trains of the Santa Fe Northwestern Railroad were hauling Jemez logs over a rail line through Guadalupe Canyon and taking them to a new sawmill in Bernalillo. It was a can-do era, and when the railroaders reached the formidable Guadalupe Box, they simply punched tunnels, now known as the Gilman Tunnels, through the rock and kept going. They eventually ended at a place called Ojitos Camp, near the grant's northern boundary.

But with timber as with minerals, high extraction costs and low prices ultimately doomed both the railroad and the logging company. The railroad died in 1941; in 1965 the timber company deeded the former land grant to the Santa Fe National Forest. New Mexico railroad historian Vernon Glover summarized the contemporary situation thus:

Today, motorists follow the old railroad route into the Cañon de San Diego Land Grant, awed by the savage geology of the Guadalupe Box. And even as physical signs of this colorful era fade, a few place names persist: Gilman, Porter, Deer Creek Landing, O'Neil Landing, Ojitos Camp, Holiday Mesa where a man named Holiday had a logging camp, and Schoolhouse Mesa, where a schoolhouse was built for children of the logging camps. (*Jemez Mountains Railroads: Santa Fe National Forest*)

All these activities, especially mining and logging, transformed the wilderness that had been the Jemez region. In the early 1900s the great conservationist Aldo Leopold (see *Black Range*) said New Mexico had two roadless areas totaling a million acres; one became the Gila Wilderness, the other was the Jemez Mountains. Today, the Gila is New Mexico's largest

Gilman Tunnel near the Rio Guadalupe

wilderness area, at 557,873 acres, while the Jemez Mountains have become a fishnet of roads.

The Jemez Mountains offer relatively little trail hiking in proportion to their size and diversity, though off-trail hiking is possible. No long-distance trails exist, and most shorter trails simply go to obvious attractions such as hot springs and waterfalls. Much of this has to do with complex landown-ership: National Forest, land grants, Indian tribal lands, mining claims, private in-holdings—and until Congress acted in 2000, the 155-square mile Baca Location No. 1 at the mountains' center, a huge chock stone of private land blocking cross-country travel. Also, topography limits hiking somewhat; the terrain is very complex, characterized by deep canyons flanked by insurmountable cliffs. And everywhere are roads—Forest Roads, logging roads, roads to cabins, etc.

The two exceptions—one, actually in combination—are the thirty-seven thousand-acre Bandelier Backcountry and the adjacent five thousand two hundred-acre Dome Wilderness, both offering excellent wilderness hiking. Most trails are easily accessed from the Bandelier National Monument Visitor Center; an interesting and very scenic two- or three-night backpack trip is the loop hike that begins at the visitor center, goes up Cañon de los Frijoles, ascends the mesa to the west, then crosses Upper Alamo Canyon en route to Yapashi Ruins. The route then descends into Capulin Canyon, following it downhill past Painted Cave. From there it heads back to the visitor center by way of lower Cañon de los Frijoles and Lower and Upper Frijoles Falls. But this route, like almost all in the Bandelier Backcountry, is dry, over extremely difficult terrain, often exposed and hot, and demands thorough preparation by hikers.

The most popular trails in the central Jemez Mountains are those along the East Fork of the Jemez River, accessible from NM 4. They lead between volcanic bluffs to scenic and interesting "boxes," waterfalls, and pools, great for water play on a hot summer afternoon.

To get to the top of Chicoma Mountain, take good dirt Forest Road 144 west from US 285 at the north end of Española. After many miles, this leads to the ridge connecting Chicoma Mountain and Polvadera Peak to the north. At the ridge's west end, a pack trail heads southeast 1.5 miles to the top. The same trail descends 1.9 miles via the east ridge of Chicoma to Forest Road 144.

For many people, their favorite memories of the Jemez Mountains are of soaking in a geothermally-heated, natural pool and gazing upward through ponderosa pines to a turquoise sky sprinkled with puffy, white clouds. The mountains encompass an undetermined number of hot springs, as some have been inaccessible to the public. The most popular ones on public land are Spence, across the Jemez River from NM 4, 1.8 miles north of Battleship Rock; McCauley (also called Battleship Rock Hot Springs), reached via a two-mile hike from either Battleship Rock or Jemez

Falls; and San Antonio, near the Rio San Antonio, reached via Forest Road 144 from NM 126 approximately six miles north-northwest of La Cueva. The hot springs—born of fire and now giving respite and release—are indeed an appropriate symbol for the Jemez Mountains.

San Miguel Mountains

The San Miguel Mountains are a small, Y-shaped cluster of summits, 3.5 x 5 miles, centered on the distinctive summit of Saint Peters Dome, said to take its name from its resemblance to the dome of Saint Peter's Basilica in Rome. The compact range is bounded on the north and east by Capulin Canyon and on the west and south by Sanchez Canyon. The summits are among the complex of peaks around the Valles Caldera, formed by eruptions millions of years preceding the caldera's collapse. Much of the range is within the 5,200-acre Dome Wilderness. Though the Dome Wilderness is New Mexico's smallest wilderness area, it adjoins the 37,000-acre wilderness of the Bandelier National Monument Backcountry. Hiking in the San Miguel Mountains usually focuses on Saint Peters Dome and Boundary Peak. Forest Road 289 connects the north and south parts of the range on the west and gives access to the Dome Wilderness.

The name San Miguel likely comes from the large Ancestral Puebloan site at the southern end of the range long known as San Miguel Ruins. To the Tewas of San Ildefonso Pueblo the San Miguel Mountains are known as the "bluebird tail mountains," while the Keresan-speakers of Cochiti Pueblo call them the "cottontail rabbit mountains." The federally protected sites and artifacts of these peoples are found throughout the mountains.

Sierra Nacimiento and San Pedro Mountains

Location: Along the western periphery of the Jemez Mountains, east of US 550 and NM 96.

Physiographic province: Southern Rocky Mountains.

Elevation range: 5,000–10,000 feet.

High point: Just over 10,600 feet at San Pedro Peaks.

Other major peaks: Nacimiento Peak, 9,791 feet; San Miguel Mountain, 9,473 feet; Pajarito Peak, 9,042 feet.

Dimensions: 10 x 50 miles.

Ecosystems: Piñon-juniper, ponderosa pine, Douglas fir, blue spruce, Engelmann spruce.

Counties: Sandoval and Rio Arriba.

Administration: Santa Fe National Forest—Cuba Ranger District; Carson National Forest—Coyote Ranger District.

Indian lands: Jemez Pueblo in the south and southwest; a small part of Zia Pueblo in the southwest.

Wilderness: San Pedro Parks Wilderness, 41,132 acres.

Getting there: From US 550 in Cuba, NM 126 provides the best access to the mountains. From this branch there are Forest Roads that lead to popular Gregorio Reservoir and trails leading into the San Pedro Parks Wilderness. Access from farther south is from NM 485 and Forest Road 376, which runs through the Gilman Tunnels. From the north, Forest Roads head south from NM 96 near Gallina.

The Sierra Nacimiento is easily seen as merely a western extension of the Jemez Mountains, but the proximity is misleading: the geological resemblance ends with both groups being formed of igneous rocks, and satellite images show a clear delineation. In the Jemez Mountains, the rocks are swarthy basalts and lavas or light-colored tuff and conglomerates, the outpouring of surface eruptions ending about a million years ago. The Sierra Nacimiento rocks are pink and gray granite, formed deep beneath the surface more than a billion years ago.

The Sierra Nacimiento's Precambrian granite foundation is conspicuous when viewed from US 550 west of Cuba or north of San Ysidro. The granite owes its present exposure at the surface to the aptly termed Laramide Revolution (about 70 million years ago). At that time, an ancient crustal block of Precambrian granite was thrust upward along a north-south fault, creating the slender range's steep western escarpment extending from San Ysidro to Gallina. The mountains' eastern slopes are much gentler, grading into the Jemez mountains.

Because the granite is relatively impermeable, the range has abundant surface water, especially as it also is the first uplift from the San Juan Basin encountered in New Mexico by east-moving storms, which yield approximately thirty-five inches of precipitation a year. The San Pedro Parks Wilderness, in the range's highest region, is characterized by moist, grassy basins—the "parks." Streams, while usually small, are ubiquitous. Indeed,

Cross-country skiing in the San Pedro Parks

at least twenty named streams originate here—perennial streams, not the intermittent streams of most New Mexico ranges—including the Rio Gallina, Rio Capulin, Rio de las Vacas, and two Rio Puercos, one running north, the other west and then south to the Rio Grande. The mountains being the birthplace of the latter Rio Puerco, a major central-New Mexico drainage that joins the Rio Grande near Bernardo, likely explains the range's name, "Birth Mountains."

Most maps label the mountains associated with the high meadows here the San Pedro Mountains, which coexist nicely with the San Pedro Parks. In these "mountains" are found the San Pedro Peaks, a curious label as a "peak-bagger" would need a GPS unit to determine which top one had ascended among these gently contoured, low-relief hills, inconspicuous in the surrounding uplands.

With abundant wildlife, plentiful trout in the numerous small streams, gentle topography, and an excellent trail system set in a cool environment, the San Pedro Parks are an outstanding destination for hikers and backpackers. The most popular trailheads are at San Gregorio Reservoir in the southwest part of the wilderness area, Las Palomas parking area in the southeast, and Resumidero Campground in the north. Because cattle grazing is among the traditional uses allowed in the wilderness, all drinking water in the wilderness must be treated. The Continental Divide Trail has been routed through the San Pedro Parks Wilderness (the Divide itself is farther west), entering the wilderness from the west via the Los Pinos Trail and exiting in the north via the Rio Capulin Trail.

The high, well-watered patchwork of forests and meadows is outstanding habitat for elk, mule deer, and black bears, among other Southern Rocky Mountain forest species. Rare is the backpacker who doesn't see at least some wildlife here, especially if watching the meadows at dawn or dusk.

The mountains' natural character has been preserved at least in part by their underwhelming mineral deposits. Copper was discovered around 1880, but the only commercial deposits were about five miles northeast of La Ventana and east of Cuba. Beginning in 1881 the district produced 7.5 million pounds of copper and seventy-five thousand ounces of silver, mostly before 1900 at the San Miguel Mine. In the Eureka Mine, copper was the replacement mineral in fossilized wood.

Coal once was mined underground in the Nacimiento foothills, but low demand closed the mines. Prospectors in the 1950s found numerous uranium occurrences—but low quality, and even lower demand, aborted mining.

Now the main activity in the Sierra Nacimiento is outdoor recreation. The quality of this resource here is high—and demand is growing.

CHAPTER THREE

Great Plains

aking up the eastern third of New Mexico, the Great Plains Province is subdivided into three sections. The Pecos Valley Section includes the terraced valleys of the Pecos and Canadian Rivers and adjacent piedmont and tablelands. The Llano Estacado Section is a region of high plains, extensions of even more extensive high plains topography in Panhandle Texas and Oklahoma.

The Raton Section includes high plains, basalt-capped mesas, and numerous volcanic structures, the only significant mountain features in this province. No true ranges exist here. Rather, the section is dominated by individual volcanic mountains within the Raton-Clayton and Ocate volcanic fields, superimposed on eroded valleys and canyons. The volcanic fields became active about 8 million years ago, and within the province are more than one hundred seventy extinct volcanoes. One of these, 8,818-foot Laughlin Peak, is the province's highest point.

Because of their isolation from other mountains, the volcanic peaks in this province reflect the ecology of the surrounding High Plains: steppe and prairie of grasses such as grama and bluestem, punctuated by scattered piñon-junipers and other small trees and shrubs. No rivers originate in this province, and while such major drainages as the Pecos, Canadian, and Dry Cimarron Rivers flow through the region, almost all other drainages are intermittent or ephemeral. No mountain wilderness areas have been designated in this province; the 15,760-acre Sabinoso Wilderness Study Area is centered on the Canadian River Canyon. The natural areas here are dominated by broad grasslands; in such a context, tall mesas and plateaus give a feeling of mountainous relief to an otherwise non-mountainous area.

Northeastern New Mexico—Great Plains

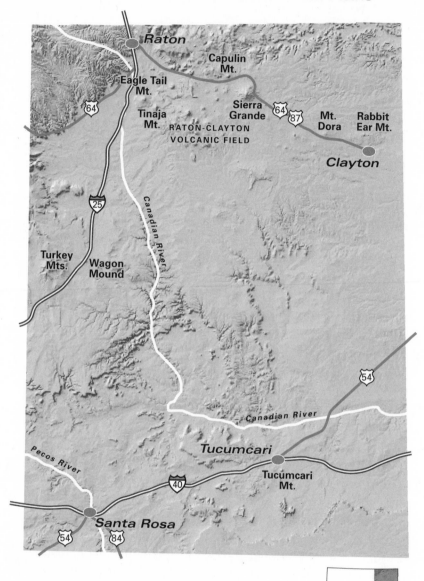

Rabbit Ear Mountain

Rabbit Ear Mountain

Location: 6 miles northwest of Clayton, east of NM 370.
Physiographic province: Great Plains—Raton Section.
Elevation: 6,058 feet.
Relief: 700 feet.
Ecosystems: Mostly mid-grass prairie, also piñon-juniper.
Counties: Union.
Administration: State and private.
Getting there: From NM 370, ranch roads branch north toward
     the mountain; landowner permission should be obtained to
     go on the private sections of the mountain.

As *The Place Names of New Mexico* explains: "While the numerous stories
behind the name of this small mountain conflict in their details, they all
agree on one thing: the mountain was named for an Indian known as
Rabbit Ear. Some accounts say he was a Cheyenne chief, others say he was
Comanche. Some say he preyed on wagon trains here. Others say he was
killed in a battle here and buried on the mountain. Most accounts say the

chief's ears had been frozen, but why this resulted in the rabbit-ear sobriquet is unknown. It's likely the name existed in Spanish as *Orejas de Conejo* before the opening of the Santa Fe Trail; certainly the mountain has been an important landmark for a long time."

Local people call the mountain "Rabbit Ears," and indeed the mountain has two summits, the smaller one 5,867 feet in elevation and one mile east of the larger. The two are perched atop Rabbit Ear Mesa, formed when basaltic lava issued from volcanic vents at the summits, approximately 3 million years ago. All are part of the Raton-Clayton Volcanic Field, where volcanic centers are aligned along fractures in the earth's crust; Rabbit Ear Mountain is at the eastern end of one such fracture.

Capulin Mountain

Location: Between Capulin and Folsom, east of NM 325, about 30 miles east of Raton, 9 miles west of Des Moines.

Physiographic province: Great Plains—Raton Section.

Elevation: 8,182 feet.

Relief: 1,200 feet.

Ecosystems: Piñon-juniper, Great Plains grasses.

Counties: Union.

Administration: National Park Service—Capulin Mountain National Monument.

Getting there: Reached by paved highway via either US 64–87 or NM 72 and then taking NM 325.

Bursting like a cyst from the surrounding plains, Capulin Mountain is the most conspicuous reminder that this part of New Mexico owes its landscape to the Raton-Clayton Volcanic Field. The region's volcanic outpourings, an extension of shifting tectonics along the Rio Grande Rift, began about 8 million years ago. At Capulin Mountain, rising gasses exploded at the surface, hurling cinders skyward; the crater's eastern rim is slightly higher than the western because of prevailing winds from the west. When the magma chamber had spent its gasses, lava began flowing, not from the main crater but from secondary vents at the volcano's eastern and western bases, allowing the cinder cone to keep its graceful symmetry. The pressure released and the volcano lapsed into long quiescence until the next magma resurgence.

The northern flanks of Capulin Mountain

Capulin Mountain is perhaps the most dramatic of numerous volcanic cinder cones in the Raton-Clayton Volcanic Field, its steep slopes of volcanic pellets having seen little erosion in the approximately fifty-six thousand years since its most recent eruption. It's this pristine symmetry that sets Capulin Mountain apart from nearby volcanoes. To showcase this, Capulin Mountain National Monument, later Capulin Volcano National Monument, was established in 1916. Today, a two-mile road coils around the cone to arrive at the crater and two short hiking trails. One, 0.2 miles, leads into the forty-five-foot-deep crater where grow the chokecherry bushes (Spanish, *capulín*) for which the peak was named. The 1.6-mile Rim Trail also is easy, and the panorama is spectacular. When I was there, the bushes along the trail were dense with ladybugs, as if the shrubs had a bumper crop of tiny orange-red berries. I was witnessing the return for mating of the convergent ladybug (*Hepatoma convergens*), a migration thousands of years old. The ladybugs begin arriving around the last week of May and continue through the first weeks of June.

From Capulin's summit are visible the cones of numerous other volcanoes. Just to the southeast is 8,720-foot Sierra Grande (see entry), a huge shield volcano made up of lava layers rather than cinder accumulations; it erupted millions of years before Capulin's eruptions. And 2.5 miles northeast of Capulin Mountain is another cinder cone named, appropriately, Baby Capulin, at 6,890 feet.

Mount Dora

Location: Immediately northeast of the village of Mount Dora, 17 miles
    northwest of Clayton, north of US 64–87.
Physiographic province: Great Plains—Raton Section.
Elevation: 6,290 feet.
Relief: 600 feet.
Ecosystems: Piñon-juniper and mid-grass prairie.
Counties: Union.
Administration: Private.
Getting there: Access is subject to landowner permission.

Mount Dora, approximately four miles in diameter, is one of several sprawling shield volcanoes in the Raton-Clayton Volcanic Field. It rises from a basalt surface studded with much smaller cinder cones and other volcanic formations. Lava at the summit dates to approximately 2.7 million years ago, while lava flowed around its base approximately 2.25 million years ago. When a nineteenth-century immigrant route ran beneath the mountain, it was called Cieneguilla del Burro Mountain, for a nearby creek. Later in the nineteenth century, Stephen W. Dorsey, U.S. Senator from Arkansas and an important landowner in northeastern New Mexico, gave the mountain its present name to commemorate his sister-in-law.

Capulin Mountain (right) and Sierra Grande (behind and left)

Sierra Grande

Location: Northeastern New Mexico, south of US 64, southeast of Raton, 4 miles southwest of Des Moines.

Physiographic province: Great Plains—Raton Section.

Elevation: 8,720 feet.

Relief: 2,100 feet.

Ecosystems: Short-grass steppe and mid-grass prairie, including grama and bluestem; piñon-juniper at higher elevations.

Counties: Union.

Administration: Mostly state, a small amount of private land.

Early local histories exaggerated grossly when they gave this mountain an elevation of 11,500 feet, but even at 8,720 feet Sierra Grande is an impressive feature—nearly eight miles in diameter and rising 2,100 feet from the plains. This explains Sierra Grande's name, "big mountain"; just to the southeast is a small subsidiary summit with the contradictory name, Little Grande.

Sierra Grande is among New Mexico's largest extinct volcanoes, and certainly the largest of the more than eighty in Union County. Some, such as Capulin Mountain, are steep-sided cinder cones, while others, such as Sierra Grande, are shield volcanoes, whose broad, gentle slopes were built up of successive lava layers. Sierra Grande was formed by multiple eruptions 3.8–2.6 million years ago. As one geologist said, "There is no reason to believe that activity has finally ceased." Perhaps Sierra Grande might yet reach 11,500 feet, though geologists say new cinder cones similar to Capulin Mountain are more likely than a resurrection of Sierra Grande.

Eagle Tail and Tinaja Mountains

Location: East of I-25, 14 miles south of Raton.

Physiographic province: Great Plains—Raton Section.

Elevation: Eagle Tail Mountain, 7,761 feet; Tinaja Mountain, 7,805 feet.

Relief: Eagle Tail Mountain, total relief (mesa bottom to top) 1,700 feet; the mountain rises about 1,350 feet atop the mesa. Tinaja Mountain, 1,300 feet.

Ecosystems: Eagle Tail Mountain, short-grass steppe and mid-grass prairie, including grama and bluestem; piñon-juniper at higher elevations. Tinaja Mountain, forested almost to the top with piñon-juniper.

Counties: Colfax.

Administration: Private.

Getting there: County Roads A6, A10, and A11 surround both mountains; further access is subject to landowner permission.

We'll likely never know who gave Eagle Tail Mountain its beautiful, descriptive name. Perhaps it was a mountain man traveling between Taos and Raton, because the resemblance to the graceful fan-shaped tail of an eagle is especially easy to see from US 64, paralleling the old mountain man trading route. Or perhaps it was an early guide or wagon master on the Santa Fe Trail, which ran just below the mountain on the west and from which the resemblance also is apparent. All we know for sure is that the name was in place by the mid-1800s, for Eagle Tail Mountain and its smaller but taller neighbor, Tinaja Mountain, four miles to the southeast, were important travelers' landmarks then, as they are now for travelers on I-25. Tinaja also has an appropriate descriptive name; it means "large earthen jar," and it takes little imagination to see the bulging sides of a large pot rising to a narrow neck.

Eagle Tail Mountain (right) and Tinaja Mountain (left)

Both mountains are the remnants of extinct volcanoes, two among more than one hundred extinct volcanoes in the Raton-Clayton Volcanic Field. Eruptions began here about 8 million years ago and continued, albeit with long quiescent periods, to Recent times, the inactive situation today likely just another lull. Eagle Tail Mountain consists of small volcanic cones and basalt layers, almost a million years old, perched atop Eagle Tail Mesa, rising 350 feet above the Canadian River floodplain and resembling a huge pancake when viewed from above. Tinaja also is a layer-cake of volcanic flows capped with a final basalt layer. While other mountains in the volcanic field are higher and larger, none bear more beautiful descriptive names.

Turkey Mountains

Location: West of Wagon Mound, south of NM 120.
Physiographic province: Great Plains—Raton Section.
Elevation range: 6,560–8,200 feet.
High point: Unnamed summit, 8,429 feet.
Dimensions: Roughly circular, diameter approximately 12 miles.
Ecosystems: Great Plains grassland, piñon-juniper, ponderosa pine.

Counties: Mora.

Administration: Private.

Getting there: Access is from the north, using dirt roads branching
    from NM 120. Access might be restricted by private ownership;
    respect all gates and private property.

For most people, the main significance of the unassuming Turkey Mountains
is that they were an important destination on the Santa Fe Trail, the Wagon
Mound Cutoff of which looped around the north and west sides of the mountains to reach Fort Union just to the southwest. The Cimarron Cutoff reached
Fort Union from the south side of the mountains. Santa Fe Trail travelers and,
later, soldiers at Fort Union found game in the mountains, including wild
turkeys. The mountains also provided firewood. Territorial Governor David
Meriwether, taking the stage east from Las Vegas in February 1854, wrote: "On
reaching a creek near what was called Barclay's Fort, a nearby trading house,
we camped with plenty of water and grass but no wood. We camped here
early in the evening because it was 15 or 20 miles to the foot of Turkey
Mountain, where we could get wood" (*My Life in the Mountains and on the
Plains: the Newly Discovered Autobiography*). He was even more eager to
reach Turkey Mountain after a snowstorm that collapsed his tent that night
and forced him to spend the night beneath layers of canvas and snow!

    The Turkey Mountains present more than a passing resemblance to an
eroded volcano, hardly surprising in the Ocate Volcanic Field, with more
than fifty cinder cones. The group is circular, with foothills that rise ever
more steeply to stop at the edge of a steep-sided central depression, and just
as one would expect from a volcano, the mountains have lava flows around
their base.

    But these are not, in fact, volcanic mountains. Rather, the igneous rock
that intruded here had solidified far below the surface, and when this mass
arose in mid-Cenozoic time, it created a dome ringed with the cracked and
tilted Permian to Cretaceous sedimentary layers it had pushed aside. The
lava came from elsewhere, but not here. The dome structure is readily seen
on topographic maps—a concentric eliptical set of cuestas, approximately
twelve miles in diameter, with steep interior walls, the highest points in the
mountains. The cuestas have four conspicuous breaches: Cottonwood
Creek on the west-southwest, Ciruela Creek on the east-northeast, Arroyo
Tierra Blanca on the southeast, and McLamar Canyon on the north.

The travelers' beacon of Tucumcari Mountain

Tucumcari Mountain

Location: Three miles south of the town of Tucumcari.

Physiographic province: Great Plains—Pecos Section.

Elevation: 4,956 feet.

Relief: 850 feet.

Ecosystems: Short-grass and mid-grass prairie.

Counties: Quay.

Administration: Private.

Getting there: From Tucumcari, a dirt road heads south from the I-40 frontage road and then crosses a canal to reach the mesa's base. All access is subject to landowner permission.

Tucumcari Mountain, despite the name, is a mesa. It's made up of Triassic, Jurassic, and Cretaceous sandstones and shales capped by a patch of white caliche limestone of the Ogallala Formation. It's not the only such mesa in the area. About 2.5 miles south of Tucumcari Mountain is Bulldog Mesa, at 4,665 feet, a similar formation; 5 miles south of Bulldog Mesa is sprawling Mesa Redonda, rising from 4,250 to 5,100 feet. Indeed, south and west of

Tucumcari are numerous steep-sided mesas like coastal islands, isolated fragments of the vast, flat Caprock overlying most of southeastern New Mexico. These outliers were created in Pliocene time, when the Canadian River broke through the Ogallala Formation, which elsewhere still forms the region's floor-flat surface.

But Tucumcari Mountain is indisputably the tallest of these outlier formations and the one that for millennia has been a travelers' landmark. It also is an obvious lookout, and that likely is the origin of the mesa's name, always a subject of curiosity. The local explanation, manufactured like "authentic Indian jewelry" primarily for tourist appeal, is that this mountain is where Cari, daughter of an Apache chief, and her lover, Tocom, ended their star-crossed romance in the time-honored manner of European-manufactured Indian legends: suicide. "Tocom! Cari!" cried the bereaved chief. Hogwash, cry I.

The name's reality is much more obscure. Perhaps the most plausible explanation is that it comes from a word, now altered beyond recognition, in a Plains Indian language meaning "lookout." But substantiation of that has proved elusive. Today the summit is used for electronic facilities, but with landowner permission a scramble to the top is worth the effort, just to see what the Indians saw long ago.

Custer Mountain

Location: West of Jal, immediately south of NM 128.
Physiographic province: Great Plains—Pecos Section.
Elevation: 3,232 feet.
Relief: 50 feet.
Ecosystems: Chihuahuan Desert scrub, including saltbush and mesquite,
    and grasses, including bluestem and grama.
Counties: Lea.
Administration: Private.
Getting there: From the north end of Jal, take NM 128 west-northwest
    8.2 miles; Custer Mountain is immediately south of the road.

The feature called Custer Mountain is a classic example of how generic terms such as *mountain, peak, cerro,* and others vary widely in their application. In any other region but the generally flat Great Plains, this blip, one-fifth mile across and only a half-mile around, would hardly merit a name, much less the label "Mountain," yet here even a relief of fifty feet confers significance, and to Plains Indians it certainly would have been an obvious landmark and lookout. The name is a mystery; it is unlikely that General George Armstrong Custer was ever near this mountain. For all anyone knows, a cowboy named Custer ran cows here.

Just 1.3 miles east-northeast is Old Baldy, another ironically named feature. It's just 3,243 feet in elevation, 1.3 miles around, again with only 50 feet of relief—not quite what one would expect from the name, and certainly no more "bald" than anything else in this treeless region.

Colorado Plateau

Centered on the Four Corners region of Arizona, Utah, Colorado, and New Mexico, the Colorado Plateau Province is characterized by mesas rather than mountains, Mount Taylor and the Chuska and Zuni Mountains notwithstanding. Here erosion has sculpted relatively unaltered sequences of sedimentary and volcanic rocks into a landscape of canyons and table-lands, mesas and buttes. The Zuni Mountains are the only major uplift mountains; other mountains such as the Chuskas and Mount Taylor have volcanic origins. Most of the state's volcanic plugs are within this province, including such landmarks as Ship Rock and Cabezon.

The province's Navajo Section is centered on the San Juan Basin and includes the Chuska Mountains and numerous volcanic features, as well as a fringe of mountains along the Colorado border west of Chama. Chromo Mountain, at 9,916 feet, on the Continental Divide northwest of Chama, is the section's highest point. The Zuni-Acoma Section includes the Zuni Mountains, volcanic features such as Cerro Alesna and Mount Taylor—at 11,301 feet the province's highest point.

Interlarded within the Colorado Plateau's sedimentary strata, especially in the San Juan Basin, are significant oil and gas and coal accumulations. The strata also contain New Mexico's most extensive vertebrate fossils, with Mesozoic and Cenozoic Eras especially well represented.

The vegetation here is typical of high, cold desert: open plains of grass and sage, open forests of piñon-juniper, ponderosa pine, and even spruce-fir at higher elevations. Though Mount Taylor exceeds eleven thousand feet, true alpine environments are not found here.

No mountain wildernesses exist here. Yet even mountains with a dense

Northwestern New Mexico—Colorado Plateau

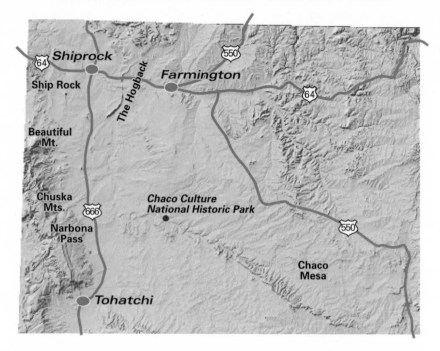

road network have retained much of their wild character. The province is conspicuously dry, and apart from brief appearances by rivers originating in Colorado, such as the San Juan, Animas, and La Plata, no significant rivers originate in this province. Nonetheless, the ephemeral streams of the province occupy impressive canyons and badland-edged valleys that hold their own wilderness appeal. Large mesas present high elevations and steep relief, invoking mountain names and respect, such as the San Mateo Mountains for Mesa Chivato.

Angel Peak

Location: 13 miles south of Bloomfield, east of US 550.
Physiographic province: Colorado Plateau—Navajo Section.
Elevation: 6,991 feet.
Relief: 600 feet.
Ecosystems: Piñon-juniper, sage, grasses.
Counties: San Juan.
Administration: BLM—Farmington Field Office.
Indian lands: Surrounded by the Navajo Nation.
Getting there: From US 550, 15 miles south of Bloomfield, County
    Road 7175 heads east 6 miles to the BLM's Angel Peak National
    Recreation Area.

Fifty million years ago, ancient rivers laid down layers of mud on floodplains, which became shale and thick layers of channel sand, which became sandstone. Within the past few million years runoff and wind erosion removed much of the surrounding shale, leaving wrinkled sandstone badlands. The top layer, called the San Jose Formation, survived to provide the crown for Angel Peak. Still, the removal of surrounding deposits that formed this peak continues, as does erosion of the Angel's wings and head, leaving us to wonder how long the angel will remain.

Huerfano Mountain

Location: 30 miles southeast of Bloomfield, north of US 550,
    near Angel Peak.
Physiographic province: Colorado Plateau—Navajo Section.

Elevation: 7,474 feet.

Relief: 500 feet.

Ecosystems: Piñon-juniper, sagebrush, grasses.

Counties: San Juan.

Administration: BLM—Farmington Field Office and Navajo Nation.

Indian lands: The formation is within land inhabited by Navajos, whose tribal lands abut the mesa.

Getting there: About 0.5 mile southeast of Huerfano Trading Post on US 550, a road branches northeast from which other roads branch toward the mesa.

As *The Place Names of New Mexico* explains, early Spanish-speakers often referred to isolated formations such as this mesa as a *huerfano,* "orphan." Ironically, this prominence, often called El Huerfano, has a close relative, a much smaller mesa called *El Huerfanito.* Both are Eocene sandstone.

But like Gobernador Knob to the north, Huerfano Mountain has a significance to Navajos that goes far beyond its size, for within the vast Navajo territory defined by the four sacred mountains, Huerfano Mountain and Gobernador Knob to the north are the two sacred "inner mountains" at the center of the *Dinetah,* the ancestral Navajo homeland. The Navajo name for Huerfano Mountain can be translated as "Holy People Encircling Mountain." Another of its Navajo names means "Turning, or Rotating, Mountain," a reference to the mesa's similar appearance when viewed from all sides. The Navajo gods once tested the Navajo people to see if they could recognize their sacred mountains when seen from above, turned from their natural positions. The Navajos' presence here proves they passed the test.

Ship Rock

Location: 9 miles southwest of the town of Shiprock, west of US 180.

Physiographic province: Colorado Plateau—Navajo Section.

Elevation: 7,178 feet.

Relief: 1,675 feet.

Ecosystems: Great Basin grasses and shrubs, including sage and saltbush.

Counties: San Juan.

Administration: Navajo Nation.

Indian lands: Within the Navajo Nation.

Ship Rock's outliers and the northeast dike

Getting there: A dirt road connecting US 180 and Tribal Road 13 passes
    just to the east of Ship Rock. Hiking around the formation's
    base is allowed, but climbing the rock is prohibited by the
    Navajo Nation.

Other volcanic necks squat sullenly upon the land; Ship Rock seems to
explode from the surrounding plains like a geyser of stone. Perceptions of
this monolith change continuously as you move around it, as light and
shadows weave and unweave each other. In the many crevices on its vertical faces nest the raptors often seen soaring on thermal air currents rising
around the peak. Like its hawks and eagles and falcons, Ship Rock soars.

Many people, when first learning that Ship Rock's Navajo name means
"Winged Rock," assume the name was inspired by this soaring quality, but

Diatremes (the volcanic plugs of the San Juan Basin)

These are the mountains of myth. How could formations with names such as Rock with Wings (Ship Rock) or Frog Rock or Beelzebub not spawn legends? Here on the Colorado Plateau, where mesas formed of sedimentary strata parallel the horizon, where reds and tans and browns grade into one another, these swarthy, gnarly intrusions pierce the Earth as if intruders from the Underworld.

Of course, that's exactly what they are. Beginning about 30 million years ago, the region around the Chuska Mountains erupted in what geologist Donald L. Barr described in his book, *Navajo Country,* as a "thunderous nightmare in the Pliocene as vent after vent erupted with vengeance. The explosive and disastrous eruption of Krakatoa in historic times is probably indicative of each of the many blowouts that occurred here in the late Cenozoic. The underlying cause of these explosive eruptions remains unknown—only the dark, ominous remnants mark their once-noisy existence."

As Barr explains,

> Though generally known as volcanic necks [or plugs], these are in reality the throats of highly explosive, gaseous, violent volcanic eruptions that rocked Navajo Country. . . . Technically they are called diatremes. . . . The diatremes consist mostly of breccias [angular fragments cemented together] containing rocks from the depths below, even from the Earth's mantle. . . . Some of

in fact the Navajos found their inspiration for the wings in the three volcanic dikes radiating outward from the central spires, equispaced like three propellers on an airplane engine.

Geologists say that Ship Rock is likely the remnant of a diatreme that erupted about 30 million years ago. What we see today is the volcano's central feeder pipe. At its formation, the pipe was 2,500 to 3,250 feet beneath the surface. Since then, the rest of the volcano and surrounding rocks have eroded away, leaving the hard, resistant core of fractured volcanic rock, or breccia, interlarded with lava veins.

Ship Rock figures prominently in Navajo mythology. Monster Slayer killed Flying Monster here. Navajos believe they once lived on Ship Rock, the material within the diatremes consists of a microbreccia, sometimes called kimberlite. These often contain tiny garnets, best found in anthills, but none has yielded diamonds such as are found in the kimberlite pipes of South Africa and elsewhere.

The diatremes have survived because they are more resistant than the surrounding rocks. Thus they rise now like the stumps of ancient cypress trees, in a swamp that went dry centuries ago.

The largest and best known of the diatremes is Ship Rock (see entry), but many others pepper the Four Corners area. Some major diatremes, ranked by relief, are:

Name	Elevation	Relief
Ship Rock	7,178 feet	1,675 feet
Zilditloi Mountain	8,574 feet	1,475 feet
Bennett Peak	6,471 feet	1,175 feet
Narbona Pass Diatreme	9,370 feet	1,000 feet
Mitten Rock	6,597 feet	900 feet
Ford Butte	6,156 feet	450 feet
Twin Buttes	8,544 feet	400 feet
Beelzebub	7,747 feet	350 feet
Frog Rock	7,382 feet	300 feet

a secure redoubt against attack, until one day lightning struck the peak, splitting it, and stranding people on the rock. (Interesting how this parallels what Acoma Indians tell of their ancestors and Enchanted Mesa.)

Because of its religious significance, Navajos have designated Ship Rock off-limits for climbing, though prior to this edict it had been climbed many times, first in 1939. It's a long, difficult climb, and the successful first ascent depended upon realizing that at one spot one must first descend before ascending to the high point. Years ago someone placed on the standard route a plaque commemorating the first climber to die on the peak. Ship Rock was the setting for Tony Hillerman's popular mystery novel, *The Fallen Man*.

Ship Rock from the south

Chuska Mountains

Location: West of US 491 (formerly US 666), southwest of Shiprock, extending into Arizona west of Sanostee.

Physiographic province: Colorado Plateau—Navajo Section.

Elevation range: 5,900–9,100 feet.

High point: Unnamed high point just south of Narbona Pass, 9,370 feet.

Other major peaks: Chuska Peak, 8,795 feet.

Dimensions: 15 x 65 miles.

Ecosystems: Piñon-juniper, rising through ponderosa pine to Douglas fir and Engelmann spruce.

Counties: San Juan, McKinley in New Mexico; Apache in Arizona.

Administration: Navajo Nation.

Indian lands: The Chuska Mountains are entirely within the Navajo Nation.

Getting there: From US 491 at Sheep Springs, NM 134 heads southwest to cross the Chuska Mountains at Narbona Pass before descending to Crystal. From the pass, Tribal Road 30 heads south. The mountains also can be reached by taking Tribal Road 13 southwest to Red Valley and then over the mountains to Lukachukai; several tribal roads branch into the mountains from here.

Among Navajos, the Chuska Mountains are the "Goods of Value" Mountains. The phrase has a specialized religious meaning among Navajos, but its general meaning has been true for about as long as humans have lived in what is now New Mexico. Certainly, even the earliest people on the Colorado Plateau took advantage of the Chuska's "goods"—cool summers, water, edible and medicinal plants, good quality chert for tools, timber for fuel and construction, and abundant game.

Recent DNA research and dendrochronology has shown that many of the timbers used to construct the "great houses" at Chaco Canyon came from the Chuska Mountains, an undertaking testifying to the high level of social organization among the Ancestral Puebloans there.

By the time Navajos arrived here, migrating from their ancestral homeland, or Dinetah, farther east, the ponderosa seedlings left in the Chuska Mountains by the Chacoans would have grown 150 feet tall. From then until the present, the Chuska Mountains have been of major importance to Navajos, whose nation encompasses the mountains.

Among Navajos, the "Goods of Value Range" includes not only the Chuska Mountains but also its subranges, the Tunicha, Lukachukai, and Carrizo Mountains, all in Arizona. Navajos see the range as a reclining male deity, with Chuska Peak as the head, Narbona Pass the neck, the Tunicha and Lukachukai Mountains the body, and the Carrizos the lower extremities.

Geologically, both the Chuska Mountains and the Defiance Plateau—separated by the narrow Black Creek Valley—are like the halves of a cracked nut, opposite sides of a monocline (upwarp) known to geologists as the Defiance Uplift, which was manifested as early as the Paleozoic Era. The mountains along this uplift separate the two great basins of Navajo Country: the Black Mesa-Holbrook Basin on the west and the San Juan Basin on the east. The Navajo novelist Irvin Morris described the Chuska Mountains and Defiance Plateau as an "archipelago of well-watered islands" (*From the Glittering World: a Navajo Story*). Two-thirds of the surface water on the Navajo Nation comes from the Defiance Plateau-Chuska Mountains region.

Nevertheless, streams are uncommon in the Chuskas, though lakes, curiously, are not—scores of them, almost all small, are perched atop the range's broad, relatively flat crest.

In Narbona Pass, the range's basic geologic structure is visible: cliffs of dark-colored volcanics resting upon older sedimentary layers. Strata such as Jurassic sandstones give color to the range's western foothills. Scattered throughout the mountains are volcanic necks and plugs, such as Frog Rock and Beelzebub near the village of Navajo. One of several names Navajos have for the Chuska Mountains means "dumping of ash on the side of the hill," a reference to Navajo mythology in which the Holy People built the mountains with layers of ash. And indeed, the mountains are literally built of layers of ash, causing the mountains' sides to have a white, ashen appearance. The name Chuska is derived from a Navajo word meaning "white-colored spruce or fir."

The Chuska Mountains have seen little mining, though some natural gas deposits exist around the Beautiful Mountain anticline at the north end of the mountains. (Yes, this broad eminence is beautiful, at least to this writer.) Rather, today as in the past, Navajos use the mountains not for minerals but for livestock grazing, medicinal herbs, timbers, hunting, fishing, farming, and ceremonies. Because the Chuskas have been an island of rich forests in an otherwise sparsely forested region, sawmills were established here as early as the 1880s. Logging intensified in the 1950s with the establishment of Navajo-owned enterprises. But by the 1990s the logging of the last old-growth forest,

Narbona Pass

Probably the most significant name in the Chuska Mountains is not of the mountains themselves but of the main pass through them. For more than 150 years, the name that had appeared on maps and in reports was Washington Pass, which connected Sheep Springs with Crystal, and almost everyone, especially Navajos, had assumed the name referred to George Washington or Washington, D.C. But around 1990, students at Navajo Community College in Ship Rock discovered the name's true origin. In 1849, the name was given at the pass by U.S. Army Lieutenant James Simpson to honor his commander, Colonel John Washington, who had just clashed with Navajos at the pass and killed their leader, Narbona, a proponent of peace who had been born at the pass. When Navajos learned of this, they initiated changing the name to honor not Washington but Narbona. The proposal had widespread support in New Mexico, so following the recommendation of the NM Geographic Names Committee, the BGN on December 10, 1992, in the final days of the 500th anniversary of Columbus's arrival in the New World, an anniversary Native Americans have said signified a renewed commitment to the preservation of their cultures, voted 4–0 to approve changing the name to Narbona Pass.

the "grandfather trees," led to conflict within the tribe over further cutting. Eventually the sawmill at Navajo was shut down, and tribal authorities took a more active role in resource management in the mountains.

The Navajo Nation has established several campgrounds in the Chuskas, most associated with the high-elevation lakes, which also give hunters access to prime elk, deer, and other game habitat. No developed hiking trails exist in the Chuskas, though tribal roads and old logging roads provide easy hiking access.

Mount Taylor

Location: Northeast of Grants, north of I-40.
Physiographic province: Colorado Plateau—Acoma-Zuni Section.
Elevation range: 6,500–11,000 feet.
High point: Mount Taylor, 11,301 feet.
Other major peaks: La Mosca, 11,053 feet.

Dimension: 23 x 47 miles.

Ecosystems: Piñon-juniper, ponderosa pine, Douglas fir, Engelmann
    spruce, aspen.

Counties: Cibola, McKinley, Sandoval.

Administration: Cibola National Forest—Mount Taylor Ranger District,
    BLM—Rio Puerco Office.

Indian lands: The Laguna and Acoma Pueblos extend to the fringes of
    the Mount Taylor complex, though these tribes and also Navajos
    consider the region as within their traditional tribal territories.

Getting there: The best access to the Mount Taylor complex is from
    Grants, where paved NM 547 ascends into the mountains before
    branching into numerous dirt Forest Service roads.

Mount Taylor is the hub of west-central New Mexico. Its huge, graceful cone
anchors the region, dominates the landscape. From its top, almost a third
of New Mexico is visible. No other mountain within fifty air miles comes
within one thousand feet of Taylor's elevation; the nearest eleven thousand-
foot summits are far away in the Jemez and Sangre de Cristo Mountains.
Had any human been here to witness it, the spectacle of Mount Taylor's
active phase would have been worldview shaping.

But the geologic forces that created the Mount Taylor Volcanic Field, as
geologists term it, are far more complex than simply eruptions. The volcanism began about 3.3 million years ago, during the late Pliocene, with
rhyolite magmas oozing from cracks along Grants Ridge. By 3 million years
ago the volcanism had shifted to the Mount Taylor area; eruptions of rhyolite lavas and tuffs began building the cone, and later, about 2.6 million
years ago, latite flowed from the crater to form the plateau now surrounding the mountain like a flared skirt. At the same time, basalt flowed from
cones subsidiary to the main activity. The cores of these cones today are
seen in more than fifty volcanic necks, or "plugs," around the perimeter of
the Mount Taylor Volcanic Field (see entry). (Note: The much more recent
lava flows of El·Malpais National Monument and along I-40 did *not* come
from Mount Taylor, which had been long extinct.)

The effect of these basaltic flows and their subsequent erosion was to
create a broad pedestal, forty-seven miles northwest to southeast and
twenty-three miles wide, ringed by a palisade of vertical columnar cliffs, a
pedestal on which rests the main crater. As Sherry Robinson, in her book *El*

Malpais, Mount Taylor, and the Zuni Mountains: a Hiking Guide and History, put it: "Taylor actually is a smallish mountain perched on a vast 8,200-foot [elevation] pedestal of mesas, which are about 2,000 feet above the surrounding mesas."

Not surprisingly, Mount Taylor has had a commanding presence in the consciousness of the peoples who have lived and traveled around it. Of the four sacred mountains that define the Navajos' universe, this is the mountain of the south, and in addition to an everyday name meaning "big, tall mountain" they refer to the peak by a ceremonial name meaning "turquoise mountain." In Navajo mythology, the peak was fastened from the sky to the earth with a great flint knife decorated with turquoise; the mountain is the home of Turquoise Boy and Yellow Corn Girl. To the Acoma Indians it is the abode of the Rainmaker of the North. Zunis call a hole on the summit "lightning hole" and believe its closure leads to drought; they once made annual summer pilgrimages to keep it open. On the mountain are sites sacred to several Indian groups.

According to *The Place Names of New Mexico,*

Early Spanish explorers called the mountain *Cebolleta,* "little onion," and it appears as *Sierra de la Zebolleta* on Bernardo Miera y Pacheco's 1778 map. Later Hispanic residents in the area called it *San Mateo.* The name Mount Taylor was given in September, 1849, by U.S. Army Lieutenant James H. Simpson, a member of Colonel John Washington's expedition into Navajo country. Simpson called it "one of the finest mountain peaks I have seen in this country," and in his journal he wrote: "This peak I have, in honor of the President of the United States, called Mount Taylor. Erecting itself high above the plain below, an object of vision at a remote distance, standing within the domain which has been so recently the theater of his sagacity and prowess, it exists, not inappropriately, an ever enduring monument to his patriotism and integrity." Though President Taylor never saw the mountain named for him, the naming nonetheless was inadvertently appropriate, as Taylor soon became the bulwark against plans by Texans to annex New Mexico, and it was through Taylor's determination, in the face of bitter southern opposition, that New Mexico remained a territory until it could become a state in its own right.

Winter on Mount Taylor

Today the mountain complex is still sometimes called the Cebolleta Mountains and also the San Mateo Mountains, doubtless causing confusion with features named Cebolla and Cebollita just to the south and with the San Mateo Mountains in Socorro County. The older names San Mateo Mountains and Cebolleta Mountains tend to refer to the entire volcanic plateau and have persisted on some maps, but they seem to be fading from popular usage, and besides, the story of all these mountains is really the story of Mount Taylor.

Today, most of the Mount Taylor complex is within the Cibola National Forest and is managed for multiple use by the Mount Taylor Ranger District. Logging hasn't occurred on the mountain since 1946. Past mining has been for coal and pumice, while some perlite mining continues today. Cattle graze throughout the mountains' extensive grasslands. But now and certainly into the future recreation is the dominant use. The numerous, scenic dirt roads are increasingly popular with mountain bikers, while in the winter the roads become cross-country ski and snowmobile routes. Campsites,

Mount Taylor Volcanic Field

Like a sow bear with her cubs, the giant, extinct volcano of Mount Taylor is surrounded by smaller volcanic offspring formations. When Mount Taylor erupted about 2.5 million years ago, the underlying magma chamber spawned not only the great central cone but also numerous subsidiary vents around the mountain. The molten igneous rock in these vents thrust into overlying sediments. With time, the softer sediments eroded away, leaving exposed the stark, dark volcanic spines. Most are named, and their vertiginous sides, often rising in hexagonal columns of basalt, make them visually compelling. The technical geological term for these formations is diatremes, but they're known colloquially as "necks," or more often "plugs," a usage reinforced by hikers who periodically go on binges of "plug-bagging." About fifty of these surround Mount Taylor, like mushrooms in a ring.

formal and informal, are numerous, though the mountains' conspicuous lack of water detracts somewhat from their appeal. Each February, the town of Grants sponsors the Mount Taylor Winter Quadrathlon, a unique endurance event in which competitors begin in Grants and go to Taylor's summit and back, a distance of forty-four miles, by bicycling, running, cross-country skiing, and snowshoeing. The "Quad," as it is known, can be done solo or as a member of one of several categories of teams.

Because the forests and meadows of Mount Taylor and its plateau are great habitat for deer, elk, and wild turkeys, the area is popular with hunters. Not with fisherman, however: no perennial lakes or streams exist on the plateau. And curiously, the region has but a handful of hiking trails. The most popular is the Gooseberry Spring Trail, No. 77, from Forest Road 193 to Mount Taylor's summit. The 3.2-mile route is steadily uphill but nowhere dauntingly steep, and it deserves its reputation as one of New Mexico's classic hikes.

The other major trail here is the Continental Divide Trail. No, the Continental Divide doesn't go through here, but the CDT does, and while much of the trail in the remote northeast section of the plateau is along seldom-traveled dirt roads, the Forest Service has constructed tread in some sections east of Grants. From the bottom of Lobo Canyon the CDT climbs to the mesa's edge and great views.

Cabezon Peak

Location: In the valley of the Rio Puerco, northwest of San Ysidro.

Elevation: 7,786 feet.

Relief: 2,000 feet.

Counties: Sandoval.

Administration: BLM—Albuquerque Field Office.

Wilderness: Cabezon Wilderness Study Area, 8,159 acres.

Getting there: Drive 19 miles northwest from the village of San Ysidro on
US 550—you'll see Cabezon Peak to the west—to a sign indicating San
Luis. Go west on the paved road, County Road 39. After 3.8 miles a road
branches left to go around the south side of the Cabezon Wilderness
Study Area, but to reach the peak itself, keep right through San Luis.
Twelve miles from US 550, when you're north of the peak, take the left,
or south fork in the road, cross a bridge, then continue until a rough
dirt road (consider walking this) heads east one mile to the peak.

While other volcanic necks exist in the valley of the upper Rio Puerco,
Cabezon is the one that commands everyone's attention. Rising two thousand
feet above the plain, looking like the blackened stump of a gigantic tree,
Cabezon is the largest of the necks in the Mount Taylor Volcanic Field. On
Cabezon's crown hikers find evidence of an Indian shrine, a reminder that
while Cabezon's local Navajo name means simply "black rock," the peak,
which is said to mark the eastern boundary of Navajo territory, also figures in
the Navajo myth of the Twin War Gods; they slew a giant whose congealed
blood is the black lava flow near Grants and whose head is Cabezon. Perhaps
early Spaniards were aware of this story, for the Spanish name *El Cabezón*
means "The Big Head." That was the name the mapmaker Bernardo Miera y
Pacheco used to label the peak on his map resulting from the 1776–77
Domínguez-Escalante expedition. When U.S. Army Lieutenant James H.
Simpson passed by in 1849, he mentioned seeing "the remarkable peak called
Cerro de la Cabeza."

The Rio Puerco valley once was home to a sizable rural population,
before overgrazing and flooding deepened arroyos and lowered the water
table. In the 1800s, the stage line connecting Santa Fe with western New
Mexico ran beneath Cabezon, and a stage stop village, now a ghost town on
private land, took the peak's name.

Today Cabezon Peak is on land administered by the BLM. Two state

Cabezon

endangered plants—Wright's fishhook cactus and grama grass cactus—have been found in the area, and likely a third, Knight's milkvetch, is here as well. Mule deer and pronghorn live in the area, and the ferruginous hawk, a candidate for federal threatened status, nests on the peak as do golden eagles and other raptors.

Cabezon's vertical sides, with dark, basaltic columns reminiscent of Devils Tower in Wyoming, are popular with rock climbers. Non-climbers can reach the top via an unmarked route that begins on the peak's southeast side, ascends a steep, narrow cleft, then snakes up the cliffs to the top. Ropes aren't needed, but the ascent requires hands-and-feet scrambling and is not for people who are skittish about heights.

Cerro Alesna

Location: 17 miles northeast of San Mateo, west-northwest of Mount Taylor.
Elevation: 8,064 feet.
Relief: 1,250 feet.
Counties: McKinley.
Administration: Private.
Indian lands: Within traditional tribal territories of the Navajo Nation and Zuni, Acoma, and Laguna Pueblos.

*Basaltic columns
on the north side of
Cerro Alesna*

Getting there: From NM 568 at the northwest end of Milan, NM 605
    branches east. Follow this for 13.8 miles past its junction with NM
    509. Continue on NM 605 until at 20.5 miles from NM 568, Forest
    Road 556 branches north, just before the village of San Mateo at
    22 miles. Follow Forest Road 556 north and northeast, into the
    Bartolome Fernandez Land Grant, until at 11 miles Alesna is
    1.5 mile east. Respect private property.

Were it not located in relatively uninhabited country and not conveniently
visible, this fang-like formation likely would be among New Mexico's most
photographed and well-known features. The rural Hispanics who have
been here for centuries named it *La Lesna,* "the awl," for its resemblance to
the narrow and sharp-pointed domestic tool, but some English-speakers
have found an apt metaphor in *Sharks Tooth.*

The northern flanks of Cerro de Nuestra Señora

Other Named Volcanic Plugs of the Mount Taylor Volcanic Field

El Tintero	7,222 feet
Cerro de Santa Clara	6,858 feet
Cerro de Guadalupe	6,844 feet
Cerro Chamisa Losa	6,890 feet
Cerro del Ojo de las Yeguas	6,834 feet
Cerro del Ojo Frio	6,611 feet
Cerro Pelon	7,026 feet
Cerro Parido	7,608 feet
Mesa Sarcio	6,906 feet
Cerro Cochino	7,072 feet
Cerro Cuate	7,019 feet
Cerro Salado	7,041 feet
Cerro Tinaja	6,775 feet
Cerrito Cochino	6,770 feet
Cerro de Nuestra Señora	7,315 feet
Cerro de Jacobo	7,437 feet
Cerro Santa Rosa	7,156 feet
Cerro Chato	6,918 feet
Cerro Vacio	7,519 feet
Cerrito Negro	7,024 feet

Zuni Mountains

Location: Southwest of Grants, south of I-40, east and north of NM 53, north of Zuni Pueblo.

Physiographic province: Colorado Plateau—Acoma-Zuni Section.

Elevation range: 6,400–9,000 feet.

High point: Mount Sedgwick, 9,256 feet.

Other major peaks: Lookout Mountain, 9,112 feet.

Dimensions: 25 x 60 miles.

Ecosystems: Gambel oak, piñon-juniper, ponderosa pine, Douglas fir, and southwest white pine, occasional blue spruce in moist locations.

Counties: Cibola.

Administration: Cibola National Forest—Mount Taylor Ranger District.

Indian lands: Zuni Pueblo includes the mountains' northwest section; on the south is the Ramah Chapter of the Navajo Nation.

Getting there: From Grants, the best access is via Zuni Canyon. From Gallup, NM 400 leads south to McGaffey, which connects with Forest Road 50. This parallels Oso Ridge along the length of the Zuni Mountains and links with NM 53.

In her book, *El Malpais, Mount Taylor, and the Zuni Mountains: a Hiking Guide and History,* Sherry Robinson writes, "... in the gentle beauty of the Zuni Mountains, my heart leads my feet. In times of pain and turmoil, this is the place that means solace." Yet she concedes that casual acquaintance with the mountains supports the public's general perception: boring. That has kept public awareness of the Zunis low, despite their easy proximity to Albuquerque, Grants, and Gallup. Great elk hunting and birding, great mountain-biking, but no place to come for peak-bagging, rock-climbing, trail-hiking, skiing, trout-fishing, and so on. The interesting and significant geology is understated. The hills are rounded rather than rugged. The Zuni Mountains are green and softly contoured; mountain connoisseurs in New Mexico value these qualities for their rarity.

The Zuni Mountains, like the Indians for which they're named, are a study in granite-tough resilience, and not just geologically. The mountains are green and welcoming today—despite centuries of human use. Indians of several tribes had been using the Zuni mountains for game, timber, fuelwood, and edible and medicinal plants long before Coronado in 1540 became the first European to enter the mountains. The deep layers of human activity are reflected in the mountains' many names. The Indians Coronado met, still living where Coronado found them in and near the mountains, explain the name Zuni thus: "Our ancestors called themselves *A'shiwi*, 'the flesh,' and their territory *Shilwon*, 'the land that produces flesh,' but the Spaniards called us Zuñi, which is their version of the Keresan *Sunyitsi* (unknown meaning)." Navajos know the mountains by names meaning "Zunis' Mountains," though other Navajo names include "many pines" for the ponderosas, and "place of wild onions," which indeed are common in the area and which have given their name to other local features, such as the nearby Rio Cebolla. As the names indicate, the Zuni Mountains have been an important resource area even for non-Zuni Indians. The Navajos of the Ramah Chapter have lived in the Zuni Mountains since before the Long Walk of 1863, when thousands of Navajos endured a forced march to a reservation at Bosque Redondo south of Fort Sumner. The Bosque Redondo experiment failed, and in 1868 Navajos were allowed to return to their homeland.

Hispanics, from settlements in foothills surrounding the mountain core, used the mountains much as the Indians had—for timber, fuel, game, wild plants, and useful minerals. They also pastured sheep and cattle here, and after the Civil War their herds were joined by those of American stockmen as well. By 1900 the mountains had been severely overgrazed.

Then came the railroad, and logging. Construction of the Atlantic and Pacific Railroad, which began in the 1880s, demanded staggering amounts of wood. Logging camps such as McGaffey, Sawyer, Malpais Springs, and Cold Springs sprouted in the Zunis; most camps were ephemeral, leaving only decaying log cabins, their sites and ruins still to be found by those willing to look for them. So also are the roadbeds of the narrow-gauge railroads that penetrated the mountains to bring out the logs. In Zuni Canyon logs careened from mountain top to canyon bottom down steep, wooden chutes. The mountains were clear-cut of their forests; old photos show a denuded landscape few people today would recognize.

The arrival of the railroad also opened eastern markets to market hunters. Wildlife in the Zunis suffered the same holocaust as elsewhere in New Mexico. Elk and desert bighorn sheep were among the species gone by around 1900. Sheep horns were found in the Zunis as late as 1931.

Then came the miners. Small copper deposits were worked at the small mining camps of Copperton and Diener near Redondo Canyon, west of Mount Sedgwick. Mining was even more localized and ephemeral than logging. Remains of both mining and logging camps are still visible, but they're fading rapidly; ours is one of the last generations to know them.

Given these centuries of human use and abuse, it would seem the Zunis have had nowhere to go but up. And indeed, the story of the mountains since the Forest Service acquired most of their land in 1940 is one of success. The ponderosas have returned, large open forests of them. The expansive meadows, such as Post Office Flat, are green and lush. Elk were reintroduced and, because of the marked habitat improvement, have thrived. In 2001 NM Game and Fish researchers found one thousand one hundred elk in the Zuni Mountains, far more than they expected—and far more than is good for the habitat. Cattle still graze here, but sheep are gone.

The Zunis attract few hikers, though this likely will change as Continental Divide Trail hikers become more aware of the El Malpais Segment alternative that goes through the Zunis, via Bonita and Zuni Canyons. The actual Continental Divide goes along Oso Ridge, a long, dry, and very difficult hike. No developed trails exist in the Zunis, though numerous Forest Roads lace the mountains and make for pleasant, easy walking, following old routes and locating historical sites. The area's ecological diversity and richness have made the Zunis renowned among birders; among the species they seek here are Bullock's oriole, American dipper, Williamson's sapsucker, Virginia's warbler, and dusky, Hammond's, and gray flycatchers. From the campground at Ojo Redondo, it's an easy 2.5-mile hike through ponderosa forest along Forest Roads to the summit of Mount Sedgwick, the Zunis' highest point. From Sedgwick's top the views of Mount Taylor are among the best of this always-scenic mountain.

F. Arch McCallum has explained that the name applied by early Spanish speakers, Sierra Madre, "Mother Range," a name used for the north-south ranges in Mexico, is appropriate for the Zuni Mountains, because the plants and animals here reflect the border between the Sierra Madre ecosystems to the south and the Rocky Mountain ecosystems to the north. Gray oak, Mexican whip-poor-will, and black-tailed rattlesnake all are near their northern limits here, while Hammond's flycatcher, red fox, common juniper, and monkshood all are in their southern reaches.

Geologically, the Zuni Mountains are a textbook illustration of an uplift formation. Imagine a whale surfacing at sea, its humped back rising above the water, waves washing against its sides. The ocean here is the extensive, mostly horizontal sedimentary formations of the Colorado Plateau, the whale

El Morro

is the mass of Precambrian granite, and the waves are the cliffs created when the granite cracked the sedimentary strata and pushed them upward. The whale surfaced during the Zuni-Defiance Uplift, part of the widespread mountain-building of the Pennsylvanian and Permian Periods, 323–245 million years ago. Captain Clarence E. Dutton, a surveyor with the U.S. Geological Survey, in 1879 wrote: "The platform named Zuni Mountains . . . is not a proper mountain range." Rather, he described it as a "broad, elongate dome almost completely ringed by high, inward facing cliffs" (Robinson, *El Malpais, Mt. Taylor, and the Zuni Mountains: a Hiking Guide and History*). The softer uplifted strata weathered into elongate valleys. The harder strata remained to become sandstone bluffs, such as El Morro. The same strata that form the scenic south-facing cliffs north of Thoreau can be found south of the Zunis—but facing north. Images from satellites show this clearly. Geologists *love* the Zuni Mountains as a geological showcase. Actually, almost everyone who knows the Zunis comes to love them.

Dowa Yalanne/Corn Mountain

Location: 2.8 miles east-southeast of Zuni Pueblo.
Physiographic province: Colorado Plateau—Acoma-Zuni Section.
Elevation: 7,235 feet.
Relief: 500 feet.
Counties: Cibola.
Administration: Zuni Pueblo.
Getting there: Several dirt roads surround the mountain's base,
    and a hiking route exists to the top, but all visitors should
    check with Zuni tribal authorities first.

For tourists, the 1.4-mile-long mesa of Dowa Yalanne, five hundred vertical feet of red and white sandstone layers, is a scenic backdrop for Zuni Pueblo, but to the people who live daily with the mesa its significance is far greater. Not only does the mesa's name "Corn Mountain" refer to the sacred staple crop of the Zuni people, but also it was here that the Zuni Indians sought refuge when Coronado and his Spanish soldiers attacked Zuni Pueblo in 1540. They would later seek safety here during the Pueblo Revolt of 1680. Like Katzimo/Enchanted Mesa for the people of Acoma Pueblo and like Dzil Ná'oodilii/Huerfano Mesa for Navajos, Dowa Yalanne/Corn Mountain is an integral landmark in the Zuni spiritual universe. The Zunis maintain shrines on the mesa's flat top, and as *The Place Names of New Mexico* explains, "A Zuni myth associates the mountain with the House of the Gods and the making of rain, lightning, and thunder; this has inspired the name sometimes used by English speakers, *Thunder Mountain.*"

Katzimo/Enchanted Mesa

Location: 2.3 miles northeast of Acoma Pueblo, or Sky City.
Physiographic province: Colorado Plateau—Acoma-Zuni Section.
Elevation: 6,643 feet.
Relief: 430 feet.

Enchanted Mesa

Counties: Cibola.
Administration: Acoma Pueblo.
Indian lands: Entirely within the Acoma Pueblo.
Getting there: For non-Acomas, the only way to view this butte is from
    Reservation Road 22 leading to the Sky City Visitor Center.

From the sandstone mesa of Sky City, atop its protecting cliffs, Acoma tour guides point northeast toward another mesa, smaller in diameter but with more precipitous cliffs, rising 430 feet above the plain below. That is Katzimo, the guides say, telling how long ago their ancestors dwelt atop it, secure against attacks, as the only way to the top was via a steep, difficult route. But one day, while most villagers were tending their fields below, a cataclysmic thunderstorm ravaged the mesa, obliterating the route and trapping some elderly women on top. They leaped to their deaths rather than starve separated from their people. It was this event that led the

Acomas to relocate to Sky City. The mesa's three names, in Keresan, Spanish, and English—*Katzimo, Mesa Encantada,* and Enchanted Mesa— all are associated with spirits related to this incident.

The mesa, layers of multicolored sandstone, is conspicuously scenic, and its cliffs have tempted climbers to attempt their own routes up the formation. To respect their ancestors, the Acomas have barred access to the mesa, though a few archaeologists made the ascent from 1897 to 1907; they found pottery, stone implements, and holes for wooden ladders.

CHAPTER FIVE

Basin and Range

▲▲▬ ▲▲▬

The mountains in this vast province are the most diverse in New Mexico, with every type of mountain-building represented: volcanoes, igneous intrusion, fault-block, folding, erosion, and even fossil reef. The province extends from I-25 east of Santa Fe south into Texas and Chihuahua, Mexico, then west into Arizona and east to the Pecos Plains. The name Basin and Range refers to the province's characteristic pattern of narrow, linear, north-south ranges alternating with broad basins; visualize the stripes on a zebra.

The Mexican Highland Section runs from Albuquerque south along both sides of the Rio Grande and includes fault-block mountains such as the Sandia and Manzano Mountains, paralleling the Rio Grande Rift on the east. The Socorro Mountains, Lemitar Mountains, and other mountains parallel the Rio Grande on the west. The uplift of all these ranges is associated with the spreading and deepening Rio Grande Rift (see entry, Chapter One), and ranges along the rift characteristically have steep escarpments facing toward the rift, gentler slopes facing away. In this section, 10,678-foot Sandia Peak is the highest point. Farther south, the San Andres and Oscura Mountains are fault-block slivers inserted between the Tularosa Basin and the Jornada del Muerto, while in the southwest the Little Hatchet, Animas, and Peloncillo Mountains likewise are bounded by large basins.

This same geological upheaval also triggered volcanism in the Mexican Highland Section, so the province has numerous volcanic mountains. These include cones along the Rio Grande, such as El Cerro Tomé, as well as the cones of the West Potrillo Mountains and Aden Crater. Even the linear mountains of the Bootheel—the Peloncillos, Animas, Big and Little Hatchets, among other ranges—have been affected by volcanism. Other

Starvation Peak

Starvation Peak at 7,042 feet, and because of its distinctive shape and intriguing name, commands the notice of travelers on I-25 between Las Vegas and Santa Fe. The chisel-shaped summit certainly was an important landmark for earlier travelers, especially those on the Santa Fe Trail. In 1855, while he was leading a military expedition to New Mexico, Major William Anderson Thornton wrote in his diary:

> En route [from Las Vegas to Santa Fe] we passed about
> 7 miles from San Miguel, the San Barnard Mountains [sic]
> about 1,500 feet high. It rises with a rapid slope for about
> 1,400 feet and then terminates with a perpendicular, in fact
> an overhanging top resembling a stupendous castle set high
> in the air neatly roofed. I could not bewish that the stars and
> stripes were wavering from its top. General Kearny strove to
> place a flag on it, when he marched around its foot during the
> war with Mexico, but he found it an impossible job.

Interestingly, Major Thornton did not refer to the mountain as Starvation Peak, a name that didn't appear in print until 1884—in an article in the *Detroit Free Press!*—about five years after the railroad eclipsed the Santa Fe Trail. Since then, the origin of the name has remained obscure. The most popular story explaining the name is that Indians ambushed travelers, presumably on the Santa Fe Trail. They sought refuge atop the peak, where the Indians blocked retreat and forced them to starve, or at least go hungry until they escaped. As *The Place Names of New Mexico* points out, the absence of evidence for the starvation incident doesn't necessarily mean it didn't occur.

mountains here, such as Salinas Peak, resulted from subsurface intrusion by igneous bodies.

To the east, the Sacramento Section includes the fault-block mass of the Sacramento Mountains, which also encompasses several volcanic structures, including 11,973-foot Sierra Blanca, the highest point of the section and the province. This section also includes the Capitan Mountains and other aligned igneous intrusions, and the Guadalupe Mountains, a fossil reef.

Within the Basin and Range Province are the classic desert ranges of the Southwest. The stark ruggedness of the mountains is accentuated by the low and sparse vegetation, exemplified by the Florida and Big Hatchet Mountains. Most of the province is within the Chihuahuan Desert, typified by plants such as mesquite, creosote bush, yucca, mountain mahogany, ocotillo, agave, and numerous species of cactus. While the province includes several large and significant wilderness areas, such as the Sandia Mountain, Manzano Mountain, Capitan, and White Mountain Wildernesses, it is the yet-to-be wildernesses, or Wilderness Study Areas, that are most important here. Low populations in this region, due primarily to lack of water and harsh desert conditions, have resulted in almost all the mountains here retaining wilderness qualities, and those wild lands managed by the BLM especially are being studied for wilderness protection: the Sierra Ladrones, Caballo, Sierra de las Uvas, Robledo, Organ, Sacramento, East and West Potrillos, Florida, Cedar, Big Hatchet, Alamo Hueco, and Peloncillo Mountains. It must be mentioned as well that three ranges and related subranges here—San Andres, Oscura, and Animas—are off-limits to the public, the first two because they are within White Sands Missile Range and the last because it is being managed for ecological research.

Pedernal Hills

Location: Approximately 12 miles south-southeast of Clines Corners.
Physiographic province: Basin and Range—Sacramento Section.
Elevation range: 6,900–7,500 feet.
High point: Pedernal Mountain, 7,565 feet.
Other major peaks: None.
Dimensions: Approximately 2 miles in diameter.
Ecosystems: Piñon-juniper, mountain mahogany, grasses and shrubs.
Counties: Torrance.
Administration: Private.
Getting there: Given permission from landowners, one would drive approximately 12 miles south-southeast of Clines Corners, then hike 2 miles west through Pintada Draw.

Southern New Mexico—Basin and Range

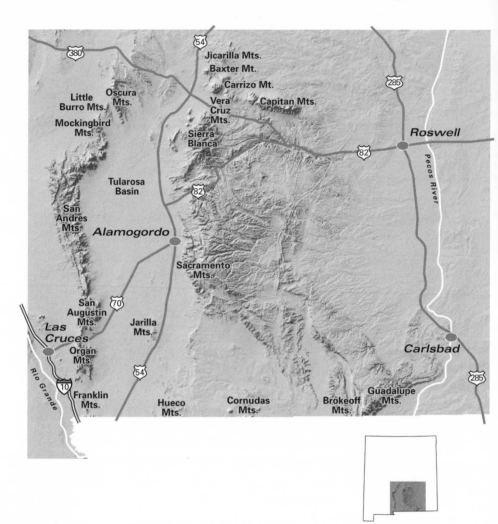

Resembling an anthill and about as memorable when glimpsed from I-40 near Clines Corners, the Pedernal Hills have a significance far beyond what their unprepossessing appearance would suggest. This small cone with a few satellite hills, the whole complex barely two miles across, is New Mexico's oldest mountain range, older even than the vast and venerable Sangre de Cristo Mountains.

In Pennsylvanian time, 318–299 million years ago, toward the end of the Paleozoic Era, mountain-building occurred in central New Mexico. North-south trending ranges stretched into Colorado, and a complex of uplifts known as the Ancestral Rocky Mountains arose. In central New Mexico, the Pedernal Uplift was the dominant feature, stretching almost to the present Texas border.

When uplift ceased, the mountains began wearing down, eventually to a flat erosional surface, mountains not again until about 200 million years later—except for the Pedernal Hills, a tough little burl of Precambrian quartzite, surrounded and then buried by much younger, sedimentary rocks. Then at the end of the Cretaceous, about 70 million years ago, the present Rocky Mountains arose during the great mountain-building of the Laramide Revolution. Of the Rockies' distant roots, only the Pedernal Hills remain. (Note: Recent geologic dating in the Pedernal Hills suggests the hills might not be as old as they appear and might have cooled as recently as 30–22 million years ago.)

The Spanish name *pedernal* would seem a misnomer as it means "flint," yet Pedernal Mountain is conspicuously and thoroughly composed of quartzite. Here, however, the quartzite often has the flaking properties of flint—and is extremely hard—so it would have been perceived and used similarly to flint. Other hills in the Pedernal Hills are granite, greenstone, and schist. The spring at the mountains' northeast base has been an important water source for travelers on the Plains since prehistoric times; a stage line once ran past it.

Gallinas Mountains

Location: West of Corona, northwest of US 54.
Physiographic province: Basin and Range—Sacramento Section.
Elevation range: 6,800–8,600 feet.
High point: Gallinas Peak, 8,637 feet.

Other major peaks: West Turkey Cone, 8,570 feet; Rough Mountain,
    8,150 feet.
Dimensions: 4.5 x 10 miles.
Ecosystems: Piñon-juniper, ponderosa pine.
Counties: Lincoln, Torrance.
Administration: Cibola National Forest—Mountainair Ranger District.
Getting there: Forest Road 167 heads west from NM 42 north of Corona,
    and Forest Road 161 heads west from US 54, 9.2 miles southwest
    of Corona—both good dirt. The roads connect, encircling the
    mountains and providing access to Forest Roads leading into
    the mountains, including Forest Road 99 through Red Cloud
    Canyon and its Forest Service campground. This road also leads
    to Forest Road 102, which goes to the top of Gallinas Peak.

The closest most people come to the Gallinas Mountains is while driving
through Corona, on their way to somewhere else. It's not that the mountains are difficult to reach; numerous good Forest Roads enter the mountains, but from a distance the mountains appear as undistinguished, low,
forested hills, a wrinkle in a carpet of piñon-juniper. But these mountains
were not always ignored.

Certainly the Puebloan Indians whose ruined villages are just north of
the mountains would have come here for game, fuelwood, and timber, just
as local people do today. Centuries later, in the late 1800s, the Gallinas
Mountains became a footnote in the annals of the Lincoln County War.
James "Whiskey Jim" Greathouse ran a stage stop and tavern here, whose
clientele ran heavily toward outlaw. In 1880, a deputy of Sheriff Pat Garrett
was killed at the tavern (Billy the Kid is rumored to have been the shooter),
and an enraged sheriff's posse burned the place to the ground.

Geologically, the Gallinas Mountains are eroded intrusions, born in
Cenozoic time when two laccoliths—a rhyolite one centered on Gallinas
Peak and a trachyte one at the north end of Rattlesnake Ridge—ascended
to the surface, elbowing aside the overlying Lower Permian sedimentary
rocks and in the process making way for iron, fluorite, and copper deposits.
Copper mining began in 1885, and by 1900 the mining camp of Red Cloud
had burgeoned around the Red Cloud Mine. Mining was most active during the 1920s, and during the 1930s fluorite mines were opened. During the
1940s, iron ore was mined, as well as the rare-earth carbonate bastnaesite.

But most mining ceased after WWII, and now few people other than locals visit the mountains.

Capitan Mountains

Location: West of Roswell, north of the Rio Bonito and US 380, northeast
    of the village of Capitan.
Physiographic province: Basin and Range—Sacramento Section.
Elevation range: 5,800–10,000 feet.
High point: Unnamed summit, 10,201 feet.
Dimensions: 10 x 22 miles.
Other major peaks: Capitan Peak, 10,083; Padilla Point, 9,627.
Ecosystems: Piñon-juniper rising through ponderosa pine to mixed
    conifers, including Douglas fir, Engelmann spruce, and corkbark fir.
Counties: Lincoln.
Administration: Lincoln National Forest—Smokey Bear Ranger District.
Wilderness: Capitan Mountains Wilderness, 35,822 acres.
Getting there: Except for a short, all-weather dirt road on their northeast,
    the Capitan Mountains are encircled by paved roads: US 380 on
    the south, NM 246 on the west and north, and NM 368 on the east.
    From these, numerous Forest Roads lead into the mountains. The
    only developed Forest Service campground is Baca, in the range's
    southern foothills.

New Mexico is the land of dry rain, waterless lakes, sand-filled streams, and rock glaciers. During the last Ice Age, about twelve thousand years ago, snow and ice accumulated on the rocky slopes of the Capitan Mountains. Ice Age conditions never were severe enough here for a true glacier to form, as on nearby Sierra Blanca, but when melt water infiltrated the rocks and then froze again, the gelid, plastic mass of ice and rock began moving downslope, like a true glacier. When the ice melted in warmer times, it left these "rock glaciers" to drape the range's slopes like dense, dark gray cobwebs. On the north side of the range, especially between Koprian and Corral Canyons, the glacial image is vivid, with tributary glaciers converging to form a medial moraine. When warmer, drier times returned about ten thousand years ago, the ice melted; the rocks were left high and dry, literally. As you hike through one of these talus fields, notice that almost all exposed rock surfaces are

Rock glacier in the Capitan Mountains

Sunset Peak in the Capitan Mountains

covered with lichens, the most patient of plants whose presence means rocks have been in their present position a long time. Without the ice, the talus slopes are remarkably stable. (See also *San Mateo Mountains*.)

But what's even more distinctive about the Capitan Mountains, though less conspicuous, is that they are one of only a few east-west trending ranges in the United States. Geology determined their orientation. During the mid-Cenozoic Era 26.5 million years ago, a magma mass about twenty miles long by four miles wide began rising from deep in the earth. As it rose, it pushed up the overlying San Andres limestone, dating from Permian time 299–251 million years ago, creating enormous pressures that were relieved along an east-west fault. The magma may never have made it to the surface to become a volcano, but it gave the rocks a distinct, buff-colored hue; it's the lichen that makes them appear dark gray. The magma eventually cooled into a solid rock mass called a pluton, and the overlying formations were eroded away; now only a few isolated patches remain, atop the highest summits.

Smokey Bear

For as long as people fight forest fires, the Capitan Mountains will be remembered as the home of New Mexico's most famous native. Not Billy the Kid—though "The Kid" knew these mountains—but rather, Smokey Bear. In 1950 a little black bear cub was found clinging, singed and burned, to a charred tree following a devastating forest fire. He became known as Smokey Bear, and generations of people worldwide came to see him as the symbol of forest fire prevention. A global survey found that only Mickey Mouse was more universally recognizable. Ironically, today many forest managers have concluded that some forest fires are necessary for natural forest succession and also to forestall catastrophic fires that result when too much fuel wood is allowed to accumulate.

Capitan Gap, a 7,542-foot pass, cleaves the range into a five-mile western segment dominated by 8,224-foot West Mountain and a seventeen-mile eastern portion with the highest tops. The range consists of a single, east-west summit ridge, incised on both sides by numerous deep, steep, and rugged canyons, often filled with rock glaciers, and equally steep ridges like ribs along a spine. Hiking here is tough, switchbacks notwithstanding, and backpacking is tougher, because the range is conspicuously lacking surface water, escept on the north side. Foothills trails paralleling the range on the north and south give access to trails ascending the ridges and canyons to the crest. The most popular hiking route is along the crest trail, accessed from the west via rough Forest Roads, and from the east via the Capitan Peak Trail from near Pine Lodge. Most summits along the crest are unnamed, including the highest point, and most are forested. The small meadow atop Capitan Peak, affording views stretching to Texas, is part of that summit's appeal for hikers. Though springs emerge along the range's base, they're conspicuously absent near the crest, except for Summit Spring.

Vegetation is typical of dry, southwestern mountains: piñon pine, alligator juniper, ponderosa pine, Gambel oak, mountain mahogany, and, along the crest, spruce and Douglas fir. Hunters know the mountains for black bear, wild turkey, deer, and elk. Hunting probably is the most common recreational use of the mountains, followed by hiking. The 35,822-acre Capitan Mountains Wilderness was designated in 1980 as part of the NM Wilderness Act.

Sacramento Mountains

Location: East of the Tularosa Basin, primarily from US 380 south to Piñon.

Physiographic province: Basin and Range—Sacramento Section.

Elevation range: 4,200–11,000 feet.

High point: Sierra Blanca, 11,973 feet.

Other major peaks: See subranges below.

Dimensions: 40 x 90 miles.

Ecosystems: Piñon-juniper, ponderosa pine, Douglas fir, aspen, spruce, and fir.

Counties: Otero, Lincoln.

Administration: Lincoln National Forest—Smokey Bear, Cloudcroft, and Mayhill Ranger Districts.

Indian lands: Mescalero Apache Reservation.

Wilderness: White Mountain Wilderness, 48,873 acres.

Getting there: See below.

As one of New Mexico's great mountain groups, the Sacramentos dominate the south as the Sangre de Cristos do the north. And of all the state's mountains, only Mount Taylor can rival Sierra Blanca as a defining regional emblem.

The Sacramento Mountains are complex, geographically and geologically, and some confusion can result from the subrange, the Sacramento Mountains south and east of Alamogordo, having the same name as the larger group. The subrange extends from US 82 south to approximately the communities of Piñon and Timberon. These mountains form the eastern boundary of the Tularosa Basin and mirror the San Andres Mountains across the basin to the west. Imagine a pie-shaped wedge taken from a watermelon, leaving two pieces of layered rind facing each other across the void.

Farther north, volcanism along the north-south zone east of the Tularosa Basin created the White Mountains, a rugged range of volcanic rocks named for its preeminent peak—Sierra Blanca.

Still farther north, north of US 380, the Sacramentos splinter into smaller ranges: south to north, the Vera Cruz (or Tucson) Mountains, Carrizo Mountain, Patos Mountain, the Jicarilla Mountains, the Tecolote Hills, and the Gallinas Mountains—all intrusive like their larger neighbors immediately south.

In historical times, the mountains have been associated with the Mescalero Apaches, who recognized an accommodating natural environment when they found one: nearly a mile higher than the surrounding arid plains, the Sacramentos offered water, cooler temperatures, game, edible and medicinal plants, and, when needed, good defensive sites. The same pleasant, livable environment draws non-Native peoples today. After years of warfare, the Mescaleros in 1883 accepted the 460,177-acre reservation, where they continue living. By skillfully using and marketing their natural resources, as well as operating a casino, the Mescaleros have prospered.

Ecologically, the Sacramento Mountains exhibit five life zones, based upon elevation:

Lower Sonoran, 3,000–4,500 feet;

Upper Sonoran, 4,500–7,500 feet;

Transition, 7,500–8,200 feet;

Northern Coniferous Forest, 8,200–10,000 feet;

Arctic-Alpine, 10,000 feet and above.

Sacramento Mountains

Location: East side of the Tularosa Basin, south of US 82.

Physiographic province: Basin and Range—Sacramento Section.

Elevation range: 4,700–9,600 feet.

High point: Alamo Peak, 9,685 feet.

Other major peaks: Sacramento Peak, 9,245 feet.

Ecosystems: Chihuahuan Desert grasses and shrubs, piñon-juniper, ponderosa pine.

Counties: Otero.

Administration: Lincoln National Forest—Cloudcroft and Mayhill Ranger Districts.

Wilderness: NM Wilderness Alliance has recommended creation of the Sacramento Escarpment Wilderness Complex totaling 34,928 acres.

Getting there: The western Sacramento Escarpment is reached from US 54 south of Alamogordo. US 82 goes north of the mountains to Cloudcroft, from which NM 130 and then NM 24 lead through the eastern foothills. Also from Cloudcroft, NM 6563 parallels the Sacramento Crest south to Sunspot.

Like so many New Mexico ranges, the Sacramento Mountains were formed when a large block, the Tularosa Basin, subsided and adjacent blocks rose— the San Andres Mountains on the west and the Sacramento Mountains on the east. Across desert flats of gypsum and mesquite, the mountains' steep escarpments face each other. In the Sacramentos, the steep western slopes rise to a crest where they meet gently inclined slopes on the east. From the Sacramento Crest, at nearly 9,700 feet, the mountains descend gradually to the Pecos River, almost one hundred miles to the east and 6,000 feet lower in elevation.

The range's rugged, west-facing escarpment has been carved from eight thousand feet of mostly Paleozoic sedimentary rocks, laid down during at least nine marine regimes. The rocks exposed here include Pennsylvanian reef formations overlain by Permian sedimentary rocks. The top layer, six hundred feet thick, is San Andres limestone, of great economic importance for it is the conduit through which water flows downhill east to produce the artesian springs that gave Artesia its name.

Incising this west-facing escarpment are deep canyons, the best known of which is Dog Canyon, about ten miles south of Alamogordo. Exposed, rough, and sometimes steep, the 4.2-mile Dog Canyon Trail nonetheless is among the state's most scenic and historically significant, for within this rugged, cliff-flanked canyon Apaches fought at least five battles with U.S. Army troops. At the canyon's mouth is Oliver Lee Memorial State Park, at the former ranch headquarters of Oliver M. Lee, a prominent and contro-versial figure in south-central New Mexico history.

Around 1900, the plans of El Paso entrepreneur Charles B. Eddy to run a railroad line up the Tularosa Basin to the promising mining camp of White Oaks had a major impact on the Sacramento Mountains. To obtain timber, a spur line, nicknamed the Cloud Climbing Railroad, was constructed and reached Cloudcroft in January 1900. Logging camps such as Russia and Marcia sprang up in the forests. Loggers cut millions of board feet in the Sacramentos for markets in Arizona and throughout the Southwest.

Though highways made the railroad obsolete (passenger service was discontinued in 1938, and the last scheduled train came down from the mountains in 1947), the railroads' impact on the Sacramentos continues today, for they made the cool, forested mountains accessible to the swel-tering citizens of Alamogordo, Las Cruces, and El Paso. In 1901 Eddy con-structed the Cloudcroft Lodge, and since then the mountains remain popular as a summer retreat. Today, travelers on US 82 pass through the Lincoln Tunnel, the longest in New Mexico.

In 1953 the National Solar Observatory, complete with "coronagraph," was constructed at the little community called Sac Peak, for nearby Sacramento Peak, but with wry humor the community changed its name to Sunspot. The mountains are famous among geologists for the world-class exposure of the Ordovician (488–444 million years ago) to the Permian (299–251 million years ago).

Hiking in the Sacramento Mountains is overshadowed by the trails in the White Mountains farther north, but the seventeen-mile Rim Trail along the Sacramento crest is ranked among the nation's top mountain-bike rides.

Sierra Blanca and the White Mountains

Location: East of the Tularosa Basin, north of US 82, south of US 380.

Physiographic province: Basin and Range—Sacramento Section.

Elevation range: 4,700–11,000 feet.

High point: Sierra Blanca, 11,973 feet.

Other major peaks: Lookout Mountain, 11,496 feet; Buck Mountain, 10,748 feet; Nogal Peak, 9,957 feet.

Dimensions: 12 x 25 miles.

Ecosystems: Piñon-juniper, alligator juniper, ponderosa pine, spruce and fir, high-elevation grasses and forbs.

Counties: Otero, Lincoln.

Administration: Lincoln National Forest—Smokey Bear Ranger District; Mescalero Apache Reservation.

Indian lands: Mescalero Apache Reservation.

Wilderness: White Mountain Wilderness, 48,873 acres.

Getting there: Forest Road 579 heading east from US 54–70 north of Alamogordo is the primary access from the west. Access from the east is much better. From NM 37 connecting Ruidoso and US 380, NM 532 heads west to Ski Apache. Forest Road 107 along the Rio Bonito and Forest Road 400 from Nogal head southwest. Access on the south is restricted by the Mescalero Apache Indian Reservation, on the north by private land.

The suffix *-est* defines Sierra Blanca. At 11,973 feet it is the highest peak in southern New Mexico, nearly 400 feet higher than its nearest rival,

Nogal Peak in the White Mountain Wilderness

11,580-foot Lookout Mountain just to the north. Sierra Blanca also can be considered New Mexico's tallest—not highest—mountain, for it has the greatest relief, rising more than seven thousand three hundred feet above the Tularosa Basin to the west. This extra reach allows Sierra Blanca to capture and retain snow longer than any other southern New Mexico summit, making it a conspicuous landmark throughout the region. This insignia also has attracted to the peak the world's most common mountain name—White Mountain—appearing in various languages on other preeminent peaks throughout the world: Mont Blanc and the Weisshorn in the Alps, Dhaulagiri in the Himalayas, Snowdon in the British Isles, Elbruz in the Caucasus, Mauna Kea in Hawaii, and Aconcagua in the Andes, among many, many others. The peak's former English name, Baldy, is common as well.

Sierra Blanca is said to be the southernmost manifestation of the Arctic life zone in the United States. Sierra Blanca also was the southernmost point in the United States to have a glacier during the last Ice Age, about twelve thousand years ago. On the mountain's northeast side is a classic glacial cirque, nine hundred feet deep and a half-mile wide, almost reaching the summit, a basin gouged by the glacier out of bedrock as if with an ice cream scoop. This cirque receives little solar warming until mid-June, allowing deep winter snow accumulations.

Sierra Blanca and the White Mountains were born in Oligocene time, 38–26 million years ago when a huge, andesitic volcano erupted here and covered about 500 square miles with 185 cubic miles of lava, ash, and volcanic fragments, for an average depth of 1,954 feet. Subsequent erosion since Miocene and Pliocene time has sculpted the mountains' slopes and canyons. Soon thereafter, a large igneous mass intruded into the volcanic deposits. Further shaping occurred with the advance of Pleistocene ice.

Because they provided life-giving water, game, and shelter, the White Mountains became an important part of the traditional territory of the Mescalero Apaches. In 1860 and again in 1873 attempts were made to settle the Mescaleros on reservations with other Apache groups as well as Navajos, but when these ended in failure the Mescaleros finally, in 1883, were given their own reservation here in the White Mountains. Since then they have skillfully managed its natural resources to develop one of the nation's most prosperous Indian reservations.

Spanish-speaking settlers from the Rio Grande arrived in the region around 1860, bestowing upon the natural features here the names by which most are still known—Sacramento, Sierra Blanca, Capitan, Jicarilla, Carrizo, Tularosa, the *rios* Hondo, Bonito, Peñasco, and more. (Sadly, not a single Apache name exists on maps of the region, despite the Apaches' long and pervasive cultural presence here.) The Hispanic settlers had not been here long before the arrival of English-speaking settlers, who with their mining, ranching, and transportation systems soon dominated the region's economic and political life.

In 1879 gold was found in Dry Gulch a few miles west of the village of Nogal, and a typical western mining boom ensued, along with the predictable demands that Indian lands be opened for mineral development. Soon, modest mining camps had sprung up—Nogal, Parsons, and Bonito. Only Nogal survives as a living community; a few foundations exist around the old Parsons mill, located in Tanbark Canyon north of Bonito Creek; Bonito now is beneath the waters of Bonito Lake.

Little did the miners suspect that the region's most valuable resource

The Crest Trail in the White Mountain Wilderness

was its climate. Beginning in 1915 when Judge Edward Medler and his wife, Lillian, of Las Cruces built the first summer cabin along the Rio Ruidoso, the proliferation of second homes, retirement homes, ski lodges and resorts, tourist shops, arts and crafts galleries, horse racing, and so forth has continued unabated. The Mescalero Apaches, astute entrepreneurs, have capitalized on this with their successful Inn of the Mountain Gods at Sierra Blanca's eastern base. Nonetheless, they protect the sanctity of their sacred peak by prohibiting access to the summit by non-Mescaleros except with their permission. Hikers out to bag New Mexico's highest peaks in the White Mountains should settle for 11,496-foot Lookout Mountain just to the north of Sierra Blanca.

Besides, the area offers great hiking elsewhere. The White Mountains Wilderness has 110 miles of trails. Most are tributaries of the twenty-seven-mile White Mountains Crest Trail, which goes through high forests and meadows from near Monjeau Lookout in the south to near Nogal Peak in the north. With excellent campsites and numerous springs, this trail is among New Mexico's finest longer trails. The 5.6-mile Three Rivers Trail, beginning at the Lincoln National Forest's Three Rivers Campground east of US 54, is the most popular trail on the White Mountains west side. More trails are found on the more heavily forested east side. Several very attractive loop hikes can be made to the crest from trailheads in Bonito Canyon. And despite the heavy tourist and recreational activity in the nearby towns, hiking pressure on the wilderness trails is surprisingly light.

Vera Cruz Mountains

Location: 10 miles southeast of Carrizozo, 1.5 miles north of US 380.

Physiographic province: Basin and Range—Sacramento Section.

Elevation range: 5,900–8,700 feet.

High point: Goat Mountain, 8,705 feet.

Other major peaks: Tucson Mountain, 8,659 feet; Vera Cruz Mountain, 7,801 feet.

Dimensions: 5 x 10 miles.

Ecosystems: Chihuahuan Desert grasses and shrubs, including mountain mahogany and yucca, piñon-juniper, alligator juniper, Gambel oak, and ponderosa pine.

Counties: Lincoln.

Administration: Lincoln National Forest—Smokey Bear Ranger District.

Getting there: Access is poor. A dirt road leads north from US 380 up
   North Indian Canyon toward Vera Cruz Mountain; jeep roads
   approach Tucson Mountain from the east.

Sometimes called the Tucson Mountains, the Vera Cruz Mountains are a
somewhat amorphous range aligned generally north-south. Vera Cruz
Mountain—steep and relatively isolated from the rest of the range—is the
most conspicuous summit, though not the highest.

The group resulted from intrusion by a small laccolith, or stock, much
smaller than the one that formed the neighboring Capitan Mountains,
Carrizo, and Patos Mountains. When the quartz monzonite intrusion
bowed the overlying Cretaceous sedimentary layers upward, mineral-
laden solutions infiltrated the interstices and formed ore bodies, here associated with breccia pipes. In 1880 a small gold-mining camp sprang up on
the west side of the Tucson Mountains, near the Vera Cruz Mine. But the
ore deposits proved to be of low grade, the camp faded, and today only
the dump of the Vera Cruz Mine remains, high on the slope of Vera Cruz
Mountain, though mining companies continue to be interested in tapping
deeper ore bodies here.

Carrizo Mountain

Like the Vera Cruz Mountains to the south and east, Carrizo Mountain was
born with the intrusion of an igneous laccolith; erosional resistance of the
laccolith compared to the surrounding sedimentary rocks resulted in the
mountain's present steep configuration. The high point is 9,605-foot Carrizo
Peak, reached by the Carrizo Peak Trail from Johnnie Canyon on the east.
Draping over the mountain's long, gentle, lower slopes are veils of talus from
the cliffs above, spreading out near the mountain's base into graceful alluvial fans, a classic portrait of desert mountain morphology. The mass of
Carrizo Mountain can be quite impressive. In the diary U.S. Army Major
William Anderson Thornton kept of the military expedition he led to New
Mexico in 1855, he wrote of seeing the "Carrizo Mountains north of northwest of us, rising high in the clouds, two miles from our camp" as he traveled toward Fort Stanton.

Although Carrizo Mountain overlooks the former mining boomtown of
White Oaks in the valley to the north, the mountain contributed little to the
area's mining activity.

Jicarilla Mountains

Location: Northeast of Carrizozo, east of US 54.

Physiographic province: Basin and Range—Sacramento Section.

Elevation range: 6,500–7,600 feet.

High point: Ancho Peak, 7,825 feet.

Other major peaks: Jicarilla Peak, 7,682 feet; Jacks Peak, 7,553 feet; Monument Peak, 7,180 feet.

Dimensions: 8 x 9.5 miles.

Ecosystems: Piñon-juniper, Gambel oak, ponderosa pine.

Counties: Lincoln.

Administration: Lincoln National Forest—Smokey Bear Ranger District.

Getting there: From White Oaks, all-weather Forest Road 72 goes through the mountains, mostly toward the east, to connect with Ancho. Numerous Forest Roads branch from this.

Though a U.S. Board on Geographic Names decision in 1982 established the Jicarilla Mountains as a group distinct from the Sacramento Mountains, many persons still consider the Jicarillas linked to the larger range, both by geography and geology. They were formed by the same tectonic episode of laccolith intrusion that produced the nearby Vera Cruz, Carrizo, and Capitan Mountains.

Gold was discovered at Baxter Mountain in 1879, and intense prospecting resulted in two mining districts: the Jicarilla District in the north and the White Oaks District in the south. By 1884, the mining boomtown of White Oaks had one thousand residents and was widely regarded as one of the territory's most promising communities. The gold deposits soon played out though the village stubbornly survives. Not all the gold is gone, however; during the Great Depression, many families lived off placer gold panned in the Jicarilla District.

The Jicarilla Mountains take their name from the Jicarilla Apaches, whose nomadic journeys would have brought them here and who once shared a reservation with the Mescalero Apaches at nearby Fort Stanton. Little hiking occurs in the mountains, primarily because of forbidding terrain and lack of water and trails.

Patos Mountain

Patos Mountain is a large ovoid mass, 2.5 by 4 miles, lying just northeast of

Carrizo Mountain and separated from it by Carrizo Canyon and White Oaks Canyon. It has several summits, aligned in an arch roughly southwest-northeast. Toward the laccolith's western end are two summits about half a mile apart, both rising to 8,490 feet. The mountain rises steadily from its base, then at about 1.2 miles its slopes abruptly become much steeper, rising approximately one thousand feet in less than a mile. Vegetation resembles that of the Capitan Mountains just to the east: piñon pine, alligator juniper, ponderosa pine, live and Gambel oak, mountain mahogany, and, at the highest elevations, spruce and Douglas fir.

In the late 1800s, freighters traveling between Socorro and Lincoln passed south of the mountain and often camped at Patos Lake. Fed by Patos Spring and Patos Creek, the lake once had plentiful water that attracted numerous wild ducks, hence the name, *patos*, "ducks." The name spread to the adjoining mountain. Except in connection with hunting, hiking in the mountains is minimal, though Forest Service Trail 75 goes over the mountain to connect the old freighter route along the mountain's south side with the Barber Springs Trail along the north side. Except for springs around the mountains' base, little water exists on the mountain.

Hueco Mountains

Location: On the New Mexico-Texas border, 20 miles east of El Paso, mostly in Texas.

Physiographic province: Basin and Range—Sacramento Section, on the boundary with Mexican Highland Section.

Elevation range: 4,300–5,400 feet.

High point: In Texas, Cerro Alto, 6,786 feet; in New Mexico, unnamed summit, 6,057 feet.

Other major peaks: None named.

Ecosystems: Chihuahuan Desert grasses and scrub, including mesquite, creosote bush, yucca, ocotillo, and cacti, with infrequent junipers.

Counties: Otero.

Administration: BLM—Las Cruces Field Office.

Getting there: From El Paso, take US 62–180 east to where a road branches north and leads to Hueco Tanks State Park. About 6 miles farther east, a maintained dirt road branches north to run along the range's east side into New Mexico.

Like the Cornudas and Franklin Mountains, the Hueco Mountains are mostly in Texas. Of the 30-mile, north-south reach of the Huecos, fewer than 7.5 miles are in New Mexico. Moreover, Texas has the well-known Hueco Tanks, now the focus of a 860-acre state historic park. For millennia travelers came here for the rainwater that collects in characteristic pockets, or *huecos,* in the igneous rock. The record is continuous since at least eleven thousand years ago; Indians of the Jornada Mogollon Culture (AD 850–1350) painted the rock surfaces with enigmatic images, including several of Tlaloc, the Mesoamerican rain god. Similar rock art is found in the New Mexico portion of the mountains, though far less known.

Such igneous formations as the Hueco Tanks are anomalous outliers of the Hueco Mountains themselves. These are primarily sedimentary formations, presenting a steep, western face of layered strata, primarily limestones deposited during Permian and Pennsylvanian time at the bottom of shallow seas. Also present is mid-Cenozoic granite, along with sills and other intrusive formations. The mountains owe their sharp relief to being tilted. The mountains are treeless, and attract few hikers, especially as the most scenic and interesting mountains are in Texas; ranching and oil and gas exploration are the main activities in the New Mexico Huecos.

Cornudas Mountains

Location: Straddling the New Mexico-Texas border, west of the Guadalupe Mountains and east of the Hueco Mountains.

Physiographic province: Basin and Range—Sacramento Section.

Elevation range: 4,800–6,000 feet.

High point: Wind Mountain, 7,280 feet.

Other major peaks: San Antonio Mountain, 6,998 feet (mostly in Texas); Cornudas Mountain, 5,730 feet.

Ecosystems: Chihuahuan Desert grasses and shrubs, including mesquite, creosote bush, yucca, ocotillo, and cacti, with infrequent junipers.

Counties: Otero.

Administration: BLM—Las Cruces Field Office, state, private.

Getting there: From US 54 north of Orogrande, NM 506 heads east to connect with maintained dirt county roads that pass south of Cornudas Mountain, eventually connecting to Dell City, TX, and then US 62–180. Numerous smaller dirt roads lead to individual formations.

Wind Mountain and locals, Cornudas Mountains

A diffuse cluster of about ten distinct laccolith formations, mid-Cenozoic in age, the Cornudas Mountains were formed when igneous masses rose to intrude into Permian limestone and other sedimentary formations. This resulted in no significant mineralization, at least not from a conventional mining sense; the feldspar-rich nepheline and phonolite rocks in the mountains, dating to the mid-Mesozoic, have been found to be perfect for melting into the brown glass used primarily for beer bottles. The atypical chemistry of the Cornudas Mountains also has led to the presence of several "oddball minerals," as geologists call them, from aergirine to xenotime.

The Cornudas Mountains are stark and barren. Indeed, on Cornudas Mountain, rugged pinnacles resembling horns caused it to it be called Horned Mountain, and likely gave the entire group its Spanish name, *cornudas.*

Wind Mountain and San Antonio Mountain are especially steep and rugged, with approximately two thousand feet of relief; they are huge cones visible on the western horizon from as far away as the Guadalupe Mountains. Cornudas Mountain, a sprawling jumble of boulders, is much lower in elevation.

Despite the forbidding terrain, Indians of the Jornada Mogollon Culture, AD 850–1350, sojourned here, camping at Persimmon and Washburn Springs at the peaks' bases, leaving incised and painted rock art similar to that found in the Alamo Hueco Mountains to the west. These same water sources also were used by stages and Pony Express riders on the Butterfield Trail, stopping at Cornudas Station at the southern base of Cornudas Mountain. (Note: Landowner permission, previously obtained, is required to visit the site.)

Steep, difficult terrain and harsh desert have discouraged hiking here; no developed trails exist. Yet despite their remoteness, the Cornudas Mountains perennially attract a few persons with a taste for solitude and desert landscapes. More recently the mountains have figured in the controversy surrounding Otero Mesa, which includes the mountains. Isolation has allowed the vast grasslands here to escape the oil and gas exploration that has dominated much of the rest of the region, but in 2005 the BLM, which owns most of the land, approved a plan opening two hundred fifty thousand acres to oil and gas development, with the possibility of more being opened later. Advocates for wilderness and natural ecology protested, pointing out that Otero Mesa, at 1.2 million acres, is North America's largest and wildest tract of Chihuahuan Desert grassland remaining in public ownership. As of this writing, the issue was hotly contested.

Brokeoff Mountains

Location: Southwest of Carlsbad, straddling the New Mexico-Texas border, northwest of and adjacent to Guadalupe Mountains National Park.

Physiographic province: Basin and Range—Sacramento Section.

Elevation range: 5,200–6,000 feet (in New Mexico).

High point: In New Mexico, 6,800 feet at the Texas border; in Texas, unnamed point, 6,950 feet.

Other major peaks: Cutoff Mountain, 6,853 feet; none named in New Mexico.

Dimensions: 6 x 18 miles.

Ecosystems: Chihuahuan Desert grasses and shrubs, including yucca, ocotillo, agaves, and cacti.

Counties: Otero.

Administration: BLM—Carlsbad Field Office, state, private.

Wilderness: The 31,606-acre BLM Brokeoff Mountains Wilderness
    Study Area is here.

Getting there: From US 285 northwest of Carlsbad, NM 137 heads
    southwest toward El Paso Gap, where a county road runs northwest
    in Big Dog Canyon between the Guadalupe and Brokeoff Mountains.
    Continuing south on NM 137 leads to the Dog Canyon campground
    and trailhead of Guadalupe Mountains National Park. Take the
    Bush Mountain Trail 3.5 miles to its junction with the Marcus Trail,
    which runs 1.2 miles past the Marcus camping area to end at the
    park border, also the Texas-New Mexico boundary, and the
    Brokeoff Mountains. Other access is via private ranch roads
    requiring landowner permission.

Separated by Dog Canyon from the main mass of the Guadalupe Mountains, the Brokeoff Mountains are less "broken off" than they are a branch, like the arm of a giant cactus. When viewed from the rim of the Guadalupes and looking west across Dog Canyon, the Brokeoffs appear as a complex of long ridges and deep canyons—like the main Guadalupes—with treeless slopes and rounded contours. Easily seen through the sparse vegetation are the limestone layers laid down during the formation of the Goat Seep Reef, predating the Capitan Reef. A hiker traveling from northwest to southeast along the range can see the transition from shallow to deeper water in the ancient sea. Now many small displacement faults have cut into the layer-cake limestones.

Despite the Brokeoffs being lower in elevation and lacking the high-elevation forests of the Guadalupes, they share similar ecology. For example, Guadalupe rabbitbrush (*Ericameria nauseosa var. texensis*), thought to exist only in the Guadalupe Mountains, also has been found in the Brokeoffs, along with such endangered plants as the Guadalupe Mountain mescal bean. Similarly, the mottled subspecies of rock rattlesnake (*Crotalus lepidus lepidus*) endangered in the Guadalupes, lives in the Brokeoffs, too. Deer and mountain lions live here, and the Brokeoffs are believed to be appropriate for reintroduction of desert bighorn sheep.

Few New Mexico mountains are as remote and seldom visited as the Brokeoffs. Private ranch roads thread through some valleys, linking stock tanks, but the mountains nonetheless remain conspicuously wild, especially the rugged ridges and canyons adjacent to Guadalupe Mountains National Park. At present, a hiker wishing to explore the wilderness of the Brokeoff Mountains would take the little-used Marcus Trail to the park's boundary and then use map and compass—and perhaps some unofficial "user" trails—to continue northward.

Guadalupe Mountains

Location: West of Carlsbad, extending south into Texas.

Physiographic province: Basin and Range—Sacramento Section.

Elevation range: 4,600–7,200 feet.

High point: Guadalupe Peak, 8,749 feet (in Texas); Camp Wilderness Ridge, 7,413 feet in New Mexico.

Other major peaks: Deer Hill, 7,056 feet; Pickett Hill, 5,450 feet.

Ecosystems: Piñon-juniper, alligator juniper, Gambel oak, mountain mahogany, ponderosa pine, Douglas fir at high elevations, as well as Chihuahuan Desert grasses and shrubs, including mesquite, creosote bush, soaptree yucca, ocotillo, and cacti.

Counties: Chaves, Eddy, Otero.

Administration: Lincoln National Forest—Guadalupe Ranger District, BLM—Carlsbad Field Office, private.

Wilderness: Carlsbad Caverns National Park Wilderness, 33,125 acres; Guadalupe Mountains National Park Wilderness (in Texas), 86,416 acres; Devils Den Canyon BLM Wilderness Study Area, 320 acres; McKittrick Canyon BLM WSA, 200 acres; Lonesome Ridge BLM WSA, 3,505 acres; Guadalupe Escarpment Forest Service WSA, 20,936 acres.

Getting there: Both Guadalupe Mountains and Carlsbad Caverns National Parks are reached from US 62–180 southwest of Carlsbad. From US 285 north of Carlsbad, NM 137 goes southwest past Sitting Bull Falls and through Queen to Forest Road 540, with access to Camp Wilderness Ridge and other ridges in the high country. NM 137 also leads to El Paso Gap, and Dog Canyon and its campground.

The huge prow of El Capitan, glowing gold at sunset as from an inner light, the peak a commanding presence over a vast and uninhabited plain—for many people that powerful image epitomizes the Guadalupe Mountains. Unfortunately for New Mexico, that image and others of the Guadalupes that are most familiar—the scarlet maples of McKittrick Canyon, the Butterfield Trail stage stop of Pine Springs, Guadalupe Mountains National Park—all are in Texas. Though more than two-thirds of the range is in New Mexico, its canyons and ridges there today are as remote and seldom visited as when Apaches came here to gather and roast mescal. Indeed, W.C. Jameson, in *The Guadalupe Mountains: Island in the Desert*, wrote: "If a Mescalero Apache of the 1870s could somehow be transported in time and returned to this Guadalupe Mountain homeland, he would find a few changes, but he would also encounter some amazing similarities to the world he departed over a hundred years earlier. Even his campsites and mescal pits remain."

But if the Guadalupe Mountains are largely ignored by the general public, they are famous among two specialized groups: geologists and cavers. The Guadalupes are part of the vast Permian Basin geological complex, from which much of the nation's oil once came, and the story of the basin and its associated reef is one of New Mexico's great geological dramas. It began at least 320 million years ago, during the Mississippian Period when the arid plains of today's southeastern New Mexico were a moderately deep ocean basin that was subsiding. On the gradient from the land to the deep ocean, shallow limestone deposits were laid down, and when the subsidence peaked about 290 to 272 million years ago, in the Permian Period, a great horseshoe-shaped reef had begun to form around the basin's periphery, reminiscent of the Great Barrier Reef off of Australia's eastern coast.

The reef here is of world-class proportions; among people who study reefs, this is a world reference. Over millennia the reef grew as lime precipitated from seawater and as countless generations of marine organisms died and deposited their calcareous skeletons; on Camp Wilderness Ridge are nubby white rocks preserving the remains of corals that lived long before the dinosaurs. Over geologic time, the reef deposits subsided, while the enclosed basin, today's Delaware Basin, filled with sediments. When the sea receded, the reef and basin were buried beneath thick layers of material eroded from nearby uplands. Erosion occurred during the early Mesozoic, but the seas returned again during the Cretaceous.

Then in late-Cenozoic time, the entire region was uplifted and tilted eastward, and as erosion removed the overlying sediments, portions of the ancient reef, now limestone, were exposed, becoming the Glass and Apache Mountains in Texas—and the Guadalupes. In a sense, all of these mountains are just one huge, horseshoe-shaped range, most of it still underground.

Deep, dry canyons in the Guadalupe Mountains

The Guadalupe Mountains in New Mexico are a range of long, deep canyons, plunging as much as two thousand feet from sprawling uplands and long, relatively level north-south ridges. From about five thousand two hundred feet elevation in the north the ridges rise to seven thousand two hundred feet in the south near the national park boundary on Camp Wilderness Ridge. On the east the mountains ascend gradually through a broad zone of foothills to end abruptly at a linear, north-south escarpment called the Rim. This overlooks Dog Canyon, which separates the Guadalupe Mountains from their northwestern extension known as the Brokeoff Mountains (see entry).

Ecologically, the Guadalupes are an "island in the desert." When the cooler, moister conditions of the Pleistocene receded about twelve thousand years ago, plants that were adapted to a cooler climate—aspens, bigtooth maples, Douglas fir, ponderosa pine, and several oak species—retreated to higher elevations. There, under conditions cooler and moister than in the desert below, they have survived, in ecological anomalies such as the Bowl high in the range. But as the climate becomes still warmer, this remnant is retreating and could disappear altogether. Other microclimates that preserve earlier flora include McKittrick Canyon and similar, almost-miraculous riparian areas and springs. These, too, are threatened by increased warming and possible accompanying drought and fires.

The large Pleistocene animals long ago vanished from the Guadalupes, victims of a changing climate and hunting by Paleo-Indians. More recently, hunting and "predator control" exterminated bighorn sheep, grizzlies, wild turkeys (later reintroduced), and wolves, and relentlessly harried black bears, mountain lions, javelinas, coyotes, and golden eagles, the latter shot from airplanes in the 1940s and 1950s. One pilot boasted of killing eight thousand golden eagles from 1945 to 1952. The creation of the national parks and other protections have allowed some species to recover.

Native Americans also are gone from the Guadalupes. Paleo-Indians, who hunted Pleistocene mammals on the southern plains, left Folsom points in Guadalupe Mountains caves, along with bones of long-extinct species. Much more recently the mountains were within the territory of the Mescalero Apaches, who came here to roast the mescal plant for which they were named. At several places in the Guadalupes their former roasting pits are visible. An 1869 military expedition broke their resistance in the Guadalupes, and by 1878 they had moved to their reservation in the White

Mountains around Sierra Blanca. They've retained their memories of the Guadalupes, however, and once a year they return to Carlsbad, where they invite people to share mescal they've roasted.

Legends have told of early Spanish explorers in these mountains, but substantiation is lacking. The most significant artifact from early Spanish days is the mountains' name, Guadalupe. In 1668 Spaniards established the Franciscan mission of *Nuestra Señora de Guadalupe* in El Paso, and, as has happened so often, the name spread by diffusion to other features. Early Spanish explorers certainly were aware of the Guadalupes, but they saw no reason to linger once their curiosity regarding precious metals was satisfied.

Certainly, the Guadalupes were unlikely candidates for mineralization, having been formed from marine limestone and sedimentary deposits. This has never deterred prospectors, treasure seekers, or legend spinners. A rich mine discovered by Coronado (who never went near the Guadalupes), pieces of gold ore captive Apaches used to buy their freedom, Geronimo proclaiming the Guadalupes the source of Apache gold, the Lost Sublett Mine, and many other tales—anyone who's beheld the mountains' pinnacles and ramparts at sunset must concede their power over the imagination. Like the Superstition Mountains of Arizona, the Guadalupes just have a mysterious aura about them, geology notwithstanding.

Of course, the Guadalupes' mystique is aided by the scores of caves within the mountains. Carlsbad Caverns has attracted worldwide attention since cowboy Jim White in 1901 saw clouds of bats emerging from the ground and discovered the cavern's entrance. More recently, Lechuguilla Cave in New Mexico, still not fully explored, has been discovered to be even more extensive. Cave specialists with the National Park Service have tallied at least 113 caves in the park. The locations of most of these are closely guarded by spelunkers fearful that inexperienced cavers will harm the extraordinarily fragile environment, not to mention themselves. To protect both caves and cavers, the locations of most caves are not made public, and iron gates bar entry except to those who have obtained a permit from Carlsbad Caverns National Park.

Most hiking in the Guadalupe Mountains occurs in Guadalupe Mountains National Park. There, trails lead from the national park visitor center and campground at Pine Springs to the top of 8,749-foot Guadalupe Peak, highest point in Texas, as well as into the backcountry. Backcountry camping is tightly regulated, with permits required. Water in the backcountry

Camp Wilderness Ridge Trail, Guadalupe Mountains

is all but non-existent. Probably the most popular day hike is the short, easy trail into McKittrick Canyon, especially in the fall, when the big tooth maples turn brilliant red. From McKittrick Canyon, the Permian Reef Trail climbs steeply to Camp Wilderness Ridge, mostly in New Mexico. A spectacular hike from New Mexico is along Camp Wilderness Ridge to this junction. Other trailheads are at the Dog Canyon trailhead and campground, approached from New Mexico. The canyons upon which these ridges look also are spectacular and interesting hiking; Upper Dark Canyon is an example.

The Guadalupe Mountains National Park Wilderness in Texas abuts Lincoln National Forest in New Mexico, which links to the 33,125-acre backcountry of Carlsbad Caverns National Park in New Mexico. The visitor center at the caverns provides access to the 13.5-mile trail along Guadalupe Ridge. The trailhead at Rattlesnake Canyon also provides backcountry access. As wilderness advocates have said, there's really only one wilderness here.

Hiking in the Guadalupes can be challenging, even dangerous—hot, waterless, exposed, over very difficult terrain—but the mountains also have a serene, gentle, almost playful nature. Anyone doubting that need only visit Sitting Bull Falls, in the foothills northwest of Carlsbad. There, amid terrain as sere and gaunt as a long-dead juniper, is a beautiful, bountiful waterfall, fed by springs flowing into emerald pools along a stream barely a mile long. Unexpected beauty—and a perfect metaphor for the Guadalupes. (Note: Another range named Guadalupe is in the extreme southwestern part of New Mexico, covered later in this chapter.)

Los Cerrillos, Ortiz Mountains, San Pedro Mountains, South Mountain

Location: South of Santa Fe, northeast of the Sandia Mountains.

Physiographic province: Basin and Range—Mexican Highland Section.

Elevation range: Los Cerrillos, 5,750–6,900 feet; Ortiz Mountains, 6,550–8,850 feet; San Pedro Mountains, 7,050–8,200 feet; South Mountain, 7,375–8,650 feet.

High point: Los Cerrillos—Cerro Bonanza, 7,088 feet; Ortiz Mountains— Placer Mountain, 8,897 feet; San Pedro Mountains—San Pedro Peak, 8,242 feet; South Mountain—unnamed point, 8,748 feet.

Other major peaks: Los Cerrillos—Grand Central Mountain, 6,976 feet; El Cerro de la Cosena, 6,923 feet; Mount Chalchihuitl, 6,310 feet; Ortiz Mountains—Lone Mountain, 7,457 feet; San Pedro— Oro Quay Peak, 8,226 feet; Cerro Columbo, 7,572 feet.

Ecosystems: Upper Sonoran grasses and shrubs, including chamisa and cacti, Gambel oak, piñon-juniper forest with scattered ponderosa pines at the highest elevations.

Counties: Santa Fe.

Administration: BLM—Taos Field Office, state, land grant, private.

Getting there: Public access is extremely limited, generally requiring landowner permission, though portions of Los Cerrillos and the San Pedro Mountains are publicly owned.

These four small mountain groups partake of a history that gives them significance far beyond their modest size. Precious minerals in these mountains at one time made them famous throughout the Southwest.

They likewise share a common geologic origin. During the mid-Cenozoic,

Albuquerque Region—Basin and Range

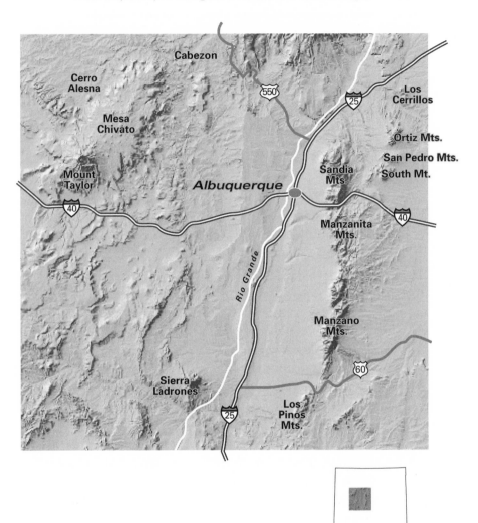

about 30 million years ago, a north-south trending fracture in the crust likely resulting from the spreading Rio Grande Rift, allowed several massive igneous bodies, called laccoliths, to begin rising toward the surface. They never emerged as volcanoes, but they did crack the overlying rocks. Pressurized mineral solutions flowed into these cracks and along rock-type boundaries, forming ore veins—turquoise, copper, tungsten, zinc, silver, and gold. Erosion left gold in placer deposits as well. If ever New Mexico possessed the bonanza sought by so many early explorers and prospectors, it was here.

Los Cerrillos

Though "The Little Hills" are not quite as small as the name implies—USGS maps label the group with the oxymoron *Cerrillos Hills*—it is true they would have scant significance were it not for their mining history, spanning more than a thousand years.

Beginning as early as AD 900, prehistoric Indians mined turquoise in Los Cerrillos, and turquoise from here is found at archaeological sites throughout the Southwest, including Chaco Canyon. Most turquoise came from Turquoise Hill, particularly Mount Chalchihuitl, whose Nahautl name meaning "blue-green stone" was given by Nahuatl-speaking Tlascalan Indian workers brought to the area by early Spaniards. As a landform, Chalchihuitl is rather underwhelming, easily overshadowed by Grand Central Mountain just to the west. Furthermore, a significant portion of Mount Chalchihuitl has been removed by quarrying. Still, its long mining history has given it a distinction beyond its physical presence and caused it to be one of few natural features to be placed on the National Register of Historic Places.

When Spaniards arrived in the sixteenth century they continued mining turquoise in Los Cerrillos, using Indians as slaves, and turquoise from here became part of the Spanish crown jewels. Though dwindling ore bodies caused active mining to cease by the late twentieth century, people perennially imagine fortunes to be made in the sacred stone, or in the tourists wanting to see its origins.

Early Spaniards mined other minerals as well in Los Cerrillos, including lead and silver, especially after the 1692 Reconquest. But the real Los Cerrillos mining boom came in 1879 when two American prospectors from Colorado discovered gold and silver deposits. The ensuing enthusiasm spawned the mining camps of Bonanza and Carbonateville and caused the otherwise sleepy village of Los Cerrillos, named for the nearby mountains,

Qui Ri Posa, *Cerrillos village cemetery*

in 1880 to have four hotels and twenty-one saloons. The boom camps were short-lived, but mining continued here longer. From 1909 to 1952 a production of 25,843 tons of metalliferous ore was obtained from 12 mines, including 920 ounces of gold, 22,187 ounces of silver, 174,894 pounds of copper, 1,379,040 pounds of lead, and 1,671,527 pounds of zinc. Coal was mined after the Atchison, Topeka and Santa Fe Railroad reached the village of Los Cerrillos in 1880, but low demand eventually closed those mines as well.

Now, perhaps for the first time in a millennium, no mining is occurring in Los Cerrillos, and the area, with significant scenic, natural, and historic values, is increasingly being used for recreation, especially by equestrians and hikers. In 1973 the Cerrillos Mining District was placed on the NM Register of Cultural Properties. Then in 2000, beginning with an initial purchase of one thousand acres, the Cerrillos Hills Historic Park was created, Santa Fe County's largest open space undertaking. Within the park the Cerrillos Hills Park Coalition has created several trails with interpretive signs highlighting some of the area's most significant mines. To reach the trailheads from the village of Cerrillos, take dirt County Road 59 north across the railroad tracks. Before reaching the cemetery, follow CR 59 left into the mouth of a dry canyon. The road here is labeled Camino Turquesa. The signed trailhead is at the canyon's mouth. Contact the Cerrillos Hills Historic Park for a schedule of guided tours.

Ortiz Mountains

The Ortiz Mountains—*mountains,* not hills—lie about five miles south of Los Cerrillos and like them owe their identity to mining. This began in early Spanish Colonial days, for the noted New Mexico mineralogist Stuart Northrop cited a reference saying the Ortiz Mine was a famous gold producer in 1680, and remnants of early Spanish mining have been reported. More modern mining was born in 1828, when a sheepherder named José Francisco Ortiz found placer gold in the mountains. Five years later vein gold was discovered, and the mining camps of Ortiz and Dolores sprang up, together known as Old Placers. By 1865 New Mexico's first stamp mill had been erected to process ores from the region's numerous lode mines, and as of 1979 an estimated seventy thousand ounces of gold had been removed from vein and placer deposits in the Ortiz area. The inventor Thomas A. Edison is reported to have spent $2 million at a laboratory in the Ortiz Mountains, trying to discover a non-water-using electrostatic method for separating gold from

*The Ortiz Mountains on the skyline, overlooking the
village of Cerrillos and the Cerrillos Cemetery*

base rock. The mountains still contain gold (and silver, lead, and tungsten), though now it is removed most economically via large open pit operations, which face formidable environmental and regulatory obstacles.

The Ortiz Mountains are a compact knot of very steep, rugged summits. The group's highest summit is, appropriately, Placer Mountain, at an elevation of 8,897 feet. As with the other groups in this chain, mining claims and land grants severely restrict public access.

San Pedro Mountains

Only Arroyo Tuerto and Arroyo la Joya separate the Ortiz and San Pedro Mountains, so it was no surprise when prospectors discovered gold in the San Pedros in 1839, soon after discoveries in the Ortiz Mountains. Gold camps— San Pedro, Golden, and Tuerto—promptly sprouted, making up the New Placers mining district. In 1846 U.S. Army Lieutenant James W. Abert visited San Pedro and found a bustling mining camp so squalid he gave thanks he didn't own a gold mine. By 1880 placer gold deposits near Golden and around the mountains' south side had yielded approximately $2 million. Gold veins also were uncovered, triggering tunnel and later open pit mining. But in addition to gold, the San Pedro Mountains have yielded other minerals, including 13 million pounds of copper since 1900. Environmental concerns have stalled large-scale mining recently, but weekend prospectors still find color in their gold pans.

Like the Ortiz Mountains, the San Pedros are rugged and compact, trending east-west three by five miles. The second highest summit, 8,226-foot Oro Quay Peak, has a name clearly referring to gold, but the name San Pedro itself likely is much older, probably derived from *San Pedro del Cuchillo,* "St. Peter of the Ridge," the seventeenth-century mission at the Pueblo of Pa'ako to the north.

South Mountain

The southernmost of the four related mountain groups oriented along a fault line—hence the name—South Mountain is the expression of an uplifted laccolith, with its northeast flank and northeast summit capped by Permian sedimentary rocks. Though prospectors seeking the same mineral wealth they had found in the northern mountains certainly scoured South Mountain as well, they found nothing comparable here. The slopes here are conspicuously steep, the upper slopes draped with talus and oak.

Sandia Mountains

Location: East of the Rio Grande, immediately east of Albuquerque.

Physiographic province: Basin and Range—Mexican Highland Section.

Elevation range: 5,800–10,000 feet.

High point: Sandia Crest, 10,678 feet.

Other major peaks: North Sandia Peak, 10,447 feet; South Sandia Peak, 9,782 feet.

Dimensions: 7 x 16 miles.

Ecosystems: Upper Sonoran desert grasses and shrubs, including bear grass, Apache plume, cholla and prickly pear cacti, and mountain mahogany; piñon-juniper at lower elevations, ponderosa pine, and near the crest, spruce, fir, and Douglas fir.

Counties: Bernalillo, Sandoval.

Administration: Cibola National Forest—Sandia Ranger District.

Indian lands: The Pueblo of Sandia includes much of the northwest portion of the mountains.

Wilderness: The 37,877-acre Sandia Mountain Wilderness straddles both sides of the Sandia crest.

A few years ago I was among several New Mexico outdoor writers asked to name their favorite mountains in the state. To my surprise, my answer was easy: the Sandias. Why the Sandias, when New Mexico has so many mountains that are higher, longer, bigger, wilder, etc.? Again, the answer was easy: they're always here when I need them.

For at least half a million New Mexicans in the Albuquerque area the Sandias are figuratively (and often literally) in their backyard. At the Elena Gallegos Open Space area in the mountains' western foothills a bronze plaque honors the conservationist Phil Tollefsrud by quoting his characterization of the mountains: "This most beautiful backdrop a city ever had." As one stands there facing the jagged granite cliffs, capped by sinuous layers of pale limestone, one is inclined to agree. Indeed, the Sandias and Albuquerque are so closely linked that Albuquerque without the Sandias would be like Denver without the Rockies. From almost anywhere in Albuquerque the mountains are no more than a thirty-minute drive away. It's where a third of New Mexico's population goes skiing, hiking, mountain-biking, rock-climbing, bow-hunting, picnicking, hang-gliding, trail-running, dog-walking, horseback riding, sightseeing, and cooling off!

The north end of the Sandia Mountains

The mountains have more than 140 miles of hiking trails, and counting. Thousands of visitors annually drive the paved road to the crest, from whence at nighttime Albuquerque appears as an ocean of light below. Not for nothing have the mountains been called an "urban wilderness." And yet. . . .

I have been aggressively exploring the Sandias since moving here twenty-five years ago; I still regularly find places completely new to me. I've backpacked along the crest and seen few other hikers, shared a high mountain meadow at night with no other campers. I know places—and not just a few—where trails are completely absent, where wild animals are common, and where Albuquerque seems not to exist. An "urban wilderness," perhaps, but a true wilderness nonetheless.

Actually, contemporary urbanites are only the most recent people to value the Sandias for just being here. In a cave overlooking Las Huertas Canyon at the north end of the mountains, archaeologist Frank Hibben found Paleo-Indian artifacts. At the time of the cave's excavation in the 1930s it was among the earliest Paleo-Indian sites in the New World. For millennia, hunter-gatherer peoples came here, and even after agricultural Puebloan culture became established along the Rio Grande, the mountains continued to provide wood, game, and edible and medicinal plants. In the foothills are the remains of tiny Puebloan farmsteads. Moreover, the Sandias continue to occupy an important place in the religious and ceremonial life of Puebloan Indians living here. To them, the Sandias are their South World Mountain. (People have been ambivalent as to whether the Sandias are a single mountain or several linked into a range.) Tiwa speakers at Sandia and Isleta Pueblos know the group as *Bien Mur*, "big mountain"; Tewas call it *Oku Pin*, "turtle mountain," for its shape. These, however, are but two of the names and identities the mountains have among contemporary Indian peoples, for the Sandias figure in all their mythologies.

When Spanish-speakers settled along the Rio Grande after 1598, they likewise used the mountains' resources, but they established no settlements because of lack of water, as well as hostile nomadic Indians. They did, however, give the mountains their present name, *Sandia*, "watermelon," most likely for the pueblo in which they'd found gourds resembling watermelons. An appealing but unsubstantiated etymology attributes the name to the mountains' resemblance to a slice of watermelon, especially when viewed from the northwest at evening, the pink granite heart capped by white limestone layers and green conifers, like a rind.

Actually, this core-rind image does describe the Sandias' geology, if not their name. The Sandia Granite at the range's core formed in the Precambrian Era and intruded older rocks 1.4 billion years ago. Immediately on top of the granite is Pennsylvanian limestone 300 million years old. Thus there is a gap of more than a billion years between the two rocks, almost a fifth of Earth's history—missing.

This billion-plus-year gap is known to geologists as the Great Unconformity. As to why no sediments remain from those billion years, two possibilities exist. First, it is possible that no sediments or other deposits were laid down during that long time. But more likely, sediments were deposited, only to be erased later by erosion.

Based upon proportions of composite minerals, Sandia Granite is technically not granite but rather grandiorite. But most geologists say the two are so closely related that even geologists refer to the rock as granite. As granites go, Sandia Granite is relatively easily eroded, which accounts for its characteristic of weathering out of the bedrock into massive rounded boulders.

Above the clouds—Kiwanis Cabin on Sandia Crest

Early Hispanics not only saw the Sandias as a chain rather than a single mountain, but they often labeled them as also including the Manzanita and Manzano Mountains to the south, separated by Tijeras Canyon. Geologically, they were correct, for both ranges share the same composition and origin. Both were created approximately 20 million years ago with faulting and stretching along the Rio Grande Rift (see *Geology,* Chapter One). By 10 million years ago, with the central part of the rift subsiding, the formations adjacent to it were uplifted, like a sidewalk section pushed upward by an expanding tree root. This process is continuing today. The result has been a steep western face backed by a more gradual eastern slope, a configuration strikingly visible from the Crest Trail, No. 130, between the Crest House and the Tram. The uplift also conferred a mile of vertical relief upon the Sandias. Actually, if one defines relief as distance from the top to the original surface, the relief is closer to three miles, because to match the limestone layer atop the Sandias with the same layer beneath the Rio Grande, one would have to dig through ten thousand feet of sediments in the Rio Grande Valley. (At Isleta, more than twenty-four thousand feet.)

Mineralization and accompanying mining was minimal in the Sandias. As with almost all other New Mexico mountains, the Sandias have in their corners the cobwebs-tales of rich mines found then later lost, but the tales are thin. The best-known tales are centered on the Placitas area, where legends speak of pre-Pueblo Revolt mines with names like *La Mina de Montezuma*. John Hayden, forester and unofficial historian of the mountains, says one story told of a man named Antonio Jiménez who set out for Mexico one day with twelve mules laden with bullion—and never was heard from again.

But far more wealth was recovered in legend than in reality, and while the areas around Tijeras and Placitas are listed among the state's mining districts, there's no evidence anything but minor mining ever occurred. Thus the mountains' wildness was left unscarred; the most conspicuous reminder of mining is the name of the mountains' most well-known trail, La Luz. In 1887, Juan Nieto and a partner opened a mine near the crest, below the present Crest House, where they worked minor deposits of fluorite, galena, lead, silver, and small amounts of gold and copper. They hauled tools and supplies up a trail that later became the old La Luz Trail. Colonel Bernard Ruppe bought the claim in 1907 and the mine briefly was called the Ruppe Mine, but it became the La Luz Mine, taking the name of the trail. Some people have said this was because the lights of Albuquerque and those of the mine cabin were mutually visible, but also the name possibly honors Nuestra Señora de la Luz, "Our Lady of the Light," one of several names for the Virgin Mary.

Hispanic settlement in the Sandias during the eighteenth and nineteenth centuries was limited to small communities in the foothills at the sites of former Indian pueblos—Placitas, Carnuel, Tijeras, Cañoncito, San Antonio, and Pa'ako—and their existence always was tenuous. The Franciscan mission at Pa'ako lasted only nine years before being abandoned in 1670, along with the pueblo it served. The Cañon de Carnuel Grant was established in 1763 to create a buffer community to protect the Rio Grande settlements from nomadic Indians, but following a devastating Apache raid in 1770 its inhabitants deserted the village, and it was not resettled until 1819. No settlements existed beyond the foothills, though livestock grazing was intense, with some landowners running as many as 2 million animals, mostly sheep, in the mountains and their foothills.

Still, for most residents of Albuquerque and other settlements along the

*The rock formation known as the Frog, off the
La Luz Trail in the Sandia Mountains*

Rio Grande, the Sandias remained forbidding and remote, a jagged line on the distant eastern horizon. That changed with the arrival of Americans and mechanization, especially after 1900. Soon after 1920, a crude road was hacked to the crest. In the 1940s, a road in Las Huertas Canyon completed a vehicular circuit around the mountains. (In 1971 a swath of forest was cleared for a proposed road north along the crest to the village of Placitas; the unpopular project was cancelled, but the swath remains.) In the 1950s, Albuquerque's population began the mushrooming growth that continues today. No longer were the Sandias distant; they were literally in people's backyards—including mine—and recreation soon dwarfed declining traditional uses of the mountains; sheep and shepherds were replaced by skiers and hikers.

Ecologically, the Sandias' mile of vertical relief means a huge diversity of ecosystems for a relatively small range. These begin with Upper Sonoran plants in the foothills: beargrass, mountain mahogany, Apache plume, prickly pear and other cacti, and scattered piñon-juniper. Above seven thousand five hundred feet the piñon-juniper gives way to ponderosa pines, in the ecological Transition Zone. Above approximately nine thousand feet is Mixed Conifer Zone, or the Canadian Zone, characterized by the same spruce and fir species one would find at a much lower elevation in British Columbia. And at the very highest elevations are other spruces and firs, as well as Douglas fir typical of the Hudsonian Zone. The next highest life zone, however, the Tundra, is not represented in the Sandias.

Wildlife in the Sandias is typical of these zones, though the wolves and grizzlies once here likely are gone forever, and endemic disease has blocked the successful reintroduction of desert bighorn sheep. The Sandias have more water sources than most desert ranges. Springs are relatively common, and perennial streams exist in Las Huertas, Domingo Baca, and Bear Canyons. None, however, flow on the surface to the Rio Grande.

Perhaps the most remarkable feature of the Sandias is how wild they've remained, given that they're surrounded by urban development. As *New Mexico's Wilderness Areas: the Complete Guide* explains:

> Today, the Sandia Mountain Wilderness [established in 1978] comprises about 45 percent of the mountains' total area. . . . Outside the wilderness, the pace of Albuquerque development has not abated. Usually only a fence separates dense housing developments from public lands. In 1981 Albuquerque purchased

Dave's Challenge, or a pillar of granite in the Sandia Mountains

portions of the former Elena Gallegos Grant in the mountains'
western foothills; 6,251 acres were added later to the Sandia
Mountain Wilderness, bringing the total wilderness acreage
to 37,877, while much of the rest was preserved as Albuquerque
Open Space, providing some buffer between city and wilderness.
In 1997, open space advocates finally succeeded in keeping
Three Gun Spring Canyon, on the mountains' south side, out
of development. In the East Mountain area, growth continues
mushrooming.

As does recreational use, whose diversity also benefits from climate and
environmental differences in the mile of vertical relief. The crest typically is
fifteen degrees Fahrenheit cooler than the foothills, and receives three times
as much precipitation annually, twenty-five to thirty inches. The main trailheads on the west are those in Juan Tabo, Embudito, and Embudo Canyons
and at the Elena Gallegos Picnic Area; on the south in Three Gun Spring
Canyon and at Canyon Estates near Tijeras; on the east via trails branching
from NM 14 and the Crest Highway, NM 536; and on the north at the Tunnel
Spring and Piedra Lisa Trailheads. The Crest Trail, No. 130, begins in the south
at the Canyon Estates trailhead and keeps close to the crest for twenty-eight
miles, ending at Tunnel Spring, passing by the Crest House en route. Among
the scores of rock-climbing routes on the cliffs of the western face, the huge
vertical granite slab known as the Shield is the best known.

Most people, however, will reach the Sandia Crest either by the Crest
Highway or the Sandia Tram, the world's longest aerial tramway, built in the
1960s to take passengers to the High Finance Restaurant at the top of the
Sandia Ski Area.

Reflecting the intense local interest in the Sandias, in 2005 the University of New Mexico Press published two books about the Sandias: *Field
Guide to the Sandia Mountains,* a unique guide to all aspects of a relatively
localized area. The book had numerous authors to deal with a broad array
of subjects, including weather, ecology, trees and shrubs, reptiles, mammals, birds, geology, human presence, and much more.

Also in 2005 UNM Press published the *Sandia Mountain Hiking Guide*
by Mike Coltrin. The book is one of the first to include not only text and
map descriptions of routes but also Global Positioning System (GPS) way-
points for important points along the trail.

Manzano Mountains

Location: South of I-40, southeast of Albuquerque, north of US 60.

Physiographic province: Basin and Range—Mexican Highland Section.

Elevation range: 6,000–10,000 feet.

High point: Manzano Peak, 10,098 feet.

Other major peaks: Gallo Peak, 10,003 feet; Bosque Peak, 9,610 feet; Mosca Peak, 9,509 feet; Guadalupe Peak, 9,450 feet; Capilla Peak, 9,360 feet; Osha Peak, 9,313 feet.

Dimensions: 15 x 46 miles.

Ecosystems: Upper Sonoran desert grasses and shrubs, including bear grass, Apache plume, cholla and prickly pear cacti, and mountain mahogany; piñon-juniper at lower elevations, ponderosa pine, and at higher elevations, spruce and Douglas fir.

Counties: Torrance, Valencia, Bernalillo.

Administration: Cibola National Forest—Tijeras and Mountainair Ranger Districts, Kirtland Air Force Base, Sandia National Laboratories.

Indian lands: Isleta Pueblo.

Wilderness: Manzano Mountain Wilderness, 36,970 acres.

Getting there: Public access to the Manzano Mountains from the north is blocked by U.S. Department of Defense withdrawals and also by Isleta Pueblo, which, along with private land, blocks most access from the west. Public access to several western trails is near the mouth of Trigo Canyon, at the former Cibola National Forest's John F. Kennedy Campground, now gated. From its intersection with NM 309 east of Belen, take NM 47 south for 5.8 miles to where a good dirt road branches east toward the mountains. Western access also is from trailheads at Monte Largo Canyon and near Encino Canyon. Access from the south is Forest Road 422 running north from US 60 west of Mountainair. The best access is from the east, by taking NM 337 south from the village of Tijeras and then NM 55 through the villages of Tajique, Torreon, and Manzano. Forest Roads head into the mountains.

Early in the twentieth century, a professional hunter spent two days tracking a mountain lion thirty miles through the Manzano Mountains; he never caught it.

Encino Canyon, on the west face of the Manzano Mountains

That incident epitomizes the Manzano Mountains. Despite their proximity to Albuquerque, the Manzanos remain in their essence wild. The
western ridges ascend steeply, with rock outcrops jagged as the teeth on a
bucksaw. It is dry, a place of gnarly, spiky plants—yucca, prickly pear cactus, scraggly piñon and junipers. On the 10,003-foot summit of Gallo Peak
are horned toads and pincushion cacti. The eastern slopes are long and
complex with canyons through which fickle streams sometimes flow, a vast
forest of piñon and juniper, ponderosa, spruce, and Douglas fir. Seen from
grassy meadows along the crest, the settlements along the Rio Grande far
below to the west seem incredibly remote; to the east the Estancia Valley
and the plains beyond dissolve in limitless distance.

To many people, the Manzano Mountains are seen as the lesser sister of
the Sandia Mountains to the north, despite the Manzanos being the larger.
Linked by the Manzanita Mountains, a transition range, the Sandia and
Manzano Mountains clearly are siblings. Indeed, early Spanish mapmakers

West face of the Manzano Mountains

sometimes labeled the two ranges as one—Sandia. Separated only by narrow Tijeras Canyon, the two ranges from a distance often appear as one. But they have important differences that go beyond map labels. Probably the most significant is human use. While the Sandias loom large in public awareness and are visited by a million persons annually—hiking, skiing, hang-gliding, driving the well-traveled paved road to the crest and its gift shop, or taking the tram to the crest and its restaurant—the Manzanos are strikingly undeveloped. The range has relatively few trailheads, and those are reached only by long drives over dirt roads; no public facilities exist on the crest, or anywhere else in the mountains, though a small observatory is maintained atop Capilla Peak. On only a few trails are hikers likely to encounter other hikers with any frequency, and many trails are traveled rarely.

The Manzano Mountains, including the Manzanita Mountains, extend approximately forty-six miles north-south. In the north, the range begins abruptly as the Manzanita Mountains rise steeply to form the south side of Tijeras Canyon, east of Albuquerque. The Manzanitas are an extension of the Manzanos, but lower and with more gentle contours. The Manzanitas originally were considered a part of the Manzano Mountains, but the U.S. Board on Geographic Names in 1982 designated them a separate subrange; still, the two clearly are related. The Manzano Mountains' southern boundary is less abrupt, the foothills extending almost to Abó Pass.

The Manzano Mountains are young mountains, at least in their most recent expression. There were uplifts here during the Laramide Orogeny about 70 million years ago, and hints that the Ancestral Rockies were here also, during the Pennsylvanian, 318–299 million years ago. Yet a mere 20 million years ago, a person looking for these ancient ranges would have seen only a plain, with no mountains at all. Then, during what geologists call the Neogene, New Mexico underwent cataclysmic geological activity that included fracturing of Earth's crust along the north-south trending Rio Grande Rift system. The resulting rift and adjacent uplifted flanks, which is continuing today, resulted in an abrupt, west-facing escarpment backed by a forest-covered slope dipping gradually to the east from the crest; picture a sidewalk slab being pushed up by a seedling growing beneath one edge of the slab. The uplift also coincided with subsidence of land immediately to the west, the resulting Albuquerque Basin in the Rio Grande Rift, a five-mile-deep, alluvium-filled trench through which the Rio Grande flows.

The western cliffs reveal the mountains' core of Precambrian granite and deformed metamorphic rocks, formed when the earth was young and little or no life existed. The granite is capped by a layer of light-colored sandstone and limestone, laid down during the Pennsylvanian Period, when this part of New Mexico was beneath the waters of a sea, at whose bottom accumulated the calcium-carbonate-rich bones and shells of uncountable sea creatures. Their fossils can be found in the limestone atop the Manzano Mountains.

Like other mountains with a basin-and-range origin, the Manzano Mountains have a linear orientation along a north-south axis. Elevations range from around 6,000 feet in the foothills to 10,098 feet at Manzano Peak, with the highest elevations in the central and south-central part of the range. Only two passes cross the range: the most important one (high point 6,515 feet) is north of Guadalupe Peak, in the Manzanita subrange. This centuries-old route between the east-Manzano villages and the Rio Grande Valley was followed by U.S. Army Lieutenant J. W. Abert during his 1846 survey of lands newly acquired from Mexico. He described it as rough. Villagers

Plateau at the northern end of the Manzano Mountains,
with Mosca and Guadalupe Peaks on the horizon

living in Chilili, Escabosa, Tajique and others remember that as children
they rode on wagons on this route through Hells Canyon. The road now is
within Isleta Pueblo and is closed to the public.

The other pass, Comanche Pass (approximately 8,300 feet), connects
Comanche Canyon on the west with Cañon de la Vereda on the east via a
trail. The pass was unnamed until 1987, when a resident of Torreon, Duncan
Simmons, proposed the name "Comanche Pass" to reflect that the pass was
along a route used by Comanche Indians traveling from the eastern plains
to raid Puebloan and Hispanic settlements along the Rio Grande.

The range takes its name from the village of Manzano in the southeast
part of the mountains; tradition says the village, in turn, took its name from
apples (Spanish: *manzanas*) grown in two ancient orchards here. These are
believed by some to have been planted by Franciscan friars in the seventeenth century, when one of the Salinas Pueblo missions was here, but the
current trees have been dated to no earlier than the early 1800s. *Manzanita
Mountains* should be read as "little Manzano Mountains," not "Little Apple

Mountains." The name the Tiwa Indians living at Isleta Pueblo have for the range means "closed fist mountains."

The Manzano Mountains are dry, though precipitation falling on the absorbent limestone atop the mountains often emerges later in the range's numerous springs. On the west side, water flowing from resistant metamorphic layers emerges as springs in Trigo Canyon and feeds the tiny stream and waterfalls there.

Lower slopes are characterized by piñon-juniper forest and then, as one goes higher, ponderosa. Because of thin, rocky soils, lack of moisture, and southern exposure, many of the upper slopes of the Manzano Mountains also include piñon-juniper forest plants such as mountain mahogany, bigtooth maple, Apache plume, yucca, and boxelder. Gambel oak and New Mexico locust also are common here, though they are successional species and not true members of this zone. Finally, as one approaches the crest, one enters mixed conifer forest, characterized by Douglas fir, white fir, blue spruce, and, where an area has been disturbed as by fire, quaking aspen.

Bigtooth maples, whose leaves turn first yellow and then scarlet in the fall, and Rocky Mountain maples with yellow leaves, are found in great concentrations in the Manzano Mountains, and when the leaves change color in late September, the Fourth of July Picnic Area in Tajique Canyon swarms with people witnessing the spectacular foliage display. Small pockets of maples exist in other Manzano canyons as well.

The size, wild character, and diversity of habitat of the Manzano Mountains result in the range being home to numerous wildlife species, including large mammals such as mountain lions, black bears, mule deer, white-tailed deer, and elk. Hawkwatch International Inc., a not-for-profit organization that monitors raptor populations, has identified a site near Capilla Peak where topography and prevailing winds funnel large numbers of eagles, hawks, falcons, kites, and other raptors past the crest as they pass on their fall migration southward. Late September through early November are the best times to see the most birds; visitors are welcome at the site.

Endangered wildlife species within the Manzano Mountains, as listed by the NM Natural Heritage Program, include the northern goshawk, the gray-footed chipmunk, Merriam's shrew, and the Mexican spotted owl.

Though little evidence remains in the Manzano Mountains of Paleo and Archaic Indians, two factors would have ensured their presence here:

one was the availability of food—big game, berries, nuts, etc.—and water in the mountains; the other was proximity to salt deposits in the evaporative lakes in the Estancia Basin immediately to the east. These also were responsible for the presence here of the Puebloan Indians.

When the Spaniards arrived in the sixteenth century, they found numerous pueblos in the Manzano Mountain foothills; these included, among others, Carnuel and Tijeras in the north, Tajique and Chilili in the east, Abó in the south, and Isleta, still inhabited on the Rio Grande to the west. Local Tiwa tradition says the present pueblo at Isleta originally was at the base of the mountains to the east. Ironically, Isleta Pueblo now bears a Spanish name, while the Hispanic villages have names based on the original Indian names for the sites.

Soon after the Spaniards' arrival, they established missions at the pueblos, but even before the Pueblo Revolt of 1680 the missions and the pueblos were abandoned, victims of raids by nomadic Indians, as well as droughts and disease. The ruins at Abó, Quarai, and Gran Quivira now make up the Salinas Pueblo Missions National Monument.

The Spaniards' lust for precious metals was disappointed in the Manzano Mountains, where mining never has been significant. Folklore says the Spaniards mined silver at a mine called *Santa Cruz,* "Holy Cross," in the southern part of the range, but that mine was lost following the Pueblo Revolt when Indians hid its location. Just north of Trigo Canyon, on the west side of the range, is a short, steep canyon called *Cañada de la Mina,* "Canyon of the Mine," whose name is unexplained, though geologists exploring north of Trigo Canyon have found veins bearing copper, gold, silver, and lead. Evidence of Spanish smelting has been found in the range's western foothills.

In the Manzanita Mountains to the north more substantial deposits have been located. In Coyote Canyon are minor lead, fluorite, copper, gold, and silver deposits, while in Hells Canyon there were at least two mines extracting gold, silver, and copper. Mining in the Manzanitas continued into the mid-twentieth century.

During the seventeenth and eighteenth centuries Hispanic people made numerous attempts at settlement in the eastern Manzano foothills, but exposure to raiding by nomadic Indians meant that most communities were short-lived. Most of the present villages date from the early-to-middle nineteenth century and the awarding of Mexican land grants, several in

*On the summit of Manzano Peak. The mailbox holds
the summit register, while the broken wooden tripod
on the right was part of the original survey of the peak.*

1841. The earliest community was Manzano, which received its land grant
in 1821, though doubtless Spanish-speaking people had been in the area
earlier. These people pursued a subsistence, land-based economy of farming, ranching, and timber and firewood cutting. Several early villages had
defensive towers, as seen in the present village of Torreon, a reminder of
pervasive Indian danger.

Today, life in these villages still has ties with this earlier economy,
though increasingly people work for wages outside their communities. In
the northeast foothills now are extensive residential areas, for people seeking a rural lifestyle to accompany employment in Albuquerque.

In 1931 what had been Manzano National Forest was redesignated the
Cibola National Forest. The Manzano Mountain Wilderness, 36,970 acres
straddling the crest, was established by the federal Endangered American
Wilderness Act of 1978.

The Manzano Mountains are rugged hiking, even by New Mexico

standards, especially the remote western escarpment. Most trails, whether from the west or the east, run to the crest and connect with the Manzano Crest Trail, which links the range's southern end with the Isleta Pueblo boundary on the north. Thus, loop trips often can be planned, ascending one trail, walking along the crest trail, then descending another. The crest trail also can be reached by driving to Capilla Peak and the observatory there.

The most popular trails on the east side include those from Fourth of July Picnic Area, where a loop hike can be made by ascending Fourth of July Canyon to the Cerro Blanco Trail 79, then continuing northwest on this to the crest for great views. Return on the Cerro Blanco Trail as it goes down a narrow canyon to meet the road, which you follow north to complete the loop.

From Red Canyon Campground another loop goes to Spruce Spring, south along the crest past the summit of Gallo Peak (it's an easy side trip to reach the summit), then down Red Canyon.

The only trailheads on the west side marked on current maps are all from the vicinity of the former John F. Kennedy Campground southeast of Belen and include the Salas, Osha, and Comanche Canyon Trails. The shortest hiking route to Manzano Peak is via the Kayser Mill Trail, reached from Forest Road 422. From there it's 2 miles to the crest, then 0.75 miles south on the Manzano Crest Trail to the peak.

The portions of the Manzano Mountains within Isleta Pueblo and Kirtland Air Force Base are closed to the public. Kirtland's closing of land around Otero Canyon off NM 337 that previously had been used by the public drew widespread opposition. Forest Service campgrounds are at Fourth of July Canyon, Capilla Peak, New Canyon, and Red Canyon.

Los Pinos Mountains

Location: Immediately south-southwest of Abó Pass, south of US 60.
Physiographic province: Basin and Range—Mexican Highland Section.
Elevation range: 5,200–7,100 feet.
High point: Whiteface Mountain, 7,530 feet.
Other major peaks: Cerro Montoso, 7,259 feet; Cerro Pelon, 6,975 feet.
Dimensions: 2.5 x 10.5 miles.
Ecosystems: Piñon-juniper, oak, mountain mahogany, saltbush and
    rabbitbrush, mixed grama grasses, cacti.

Counties: Socorro.

Administration: Almost entirely within the Sevilleta National Wildlife
   Refuge; access is restricted.

Wilderness: The entire 228,000-acre refuge has been designated a Long
   Term Ecological Research Site by the National Science Foundation
   and the U.S. Fish and Wildlife Service. For more information, contact
   the Sevilleta National Wildlife Refuge visitor center, just west of I-25
   south of Bernardo.

Getting there: Only the BLM lands on the range's eastern foothills are
   open to the public.

From I-25 the Los Pinos Mountains, especially their western escarpment, present themselves as a dense knot of gnarly rock. Their eastern slopes, due to the mountains being an uplifted fault block, are somewhat gentler, but only by comparison. This is very rugged country. Despite the mountains' name, pines, while common, are far outnumbered by junipers here—and even they are widely spaced. Thus, the vegetation is predominantly grasses, such as grama and bluestem, and Chihuahuan Desert scrub, with small cacti common.

The Los Pinos Mountains owe their ruggedness at the northern part of the range to their core of Precambrian quartzite—an extremely resistant rock. These light-gray formations are visible from US 60 west of Abó Pass. To the south, Precambrian granite, metarhyolite, and Pennsylvanian limestone are the most resistant layers.

The Los Pinos Mountains are just south of Abó Pass, where for at least eleven thousand five hundred years humans have passed through this natural corridor. The Spanish explorers found Tompiro Indians living at several sites near the pass, including the one nearby that gave its name to the pass. Their culture withered and died in the face of drought, Spanish misrule, and attacks by nomadic Indians. Their rock paintings in Abó Canyon have survived.

Little San Pascual Mountains

Location: South of San Antonio, at the southeast part of Bosque del
   Apache National Wildlife Refuge, east of San Marcial.

Physiographic province: Basin and Range—Mexican Highland Section.

Elevation range: 4,700–5,400 feet.

High point: Little San Pascual Mountain, 5,525 feet.

Dimensions: 1 x 3.3 miles.

Ecosystems: Chihuahuan Desert grasses and shrubs.

Counties: Socorro.

Administration: U.S. Fish and Wildlife Service—Bosque del Apache National Wildlife Refuge.

Wilderness: The 19,859-acre Little San Pascual Wilderness encompasses these mountains, which are almost entirely within the wildlife refuge.

Getting there: Day hiking is allowed, but camping requires advance permission, usually granted only to educational groups. Check at the Bosque del Apache National Wildlife Refuge for current access and restrictions.

Like the Chupadera Mountains to the west, the Little San Pascual Mountains are a brown bracket beside the green wetlands of Bosque del Apache. The diminutive range is a relatively low but sharp ridge of east tilted Paleozoic rocks, the same ones found elsewhere in other ranges along the east side of the Rio Grande. The Little San Pasqual Mountains are part of a much larger area under long-term ecological study. Because of this, as well as small size, poor access, and lack of water, recreational hiking in the mountains is extremely limited. Besides, there are many desert ranges along the Rio Grande—but only one Bosque del Apache.

Oscura Mountains

Location: At the north end of the San Andres Mountains, between the Jornada del Muerto and the Tularosa Basin, north of Mockingbird Gap, south of US 380.

Physiographic province: Basin and Range—Mexican Highland Section.

Elevation range: 4,900–8,200 feet.

High point: Oscura Peak, 8,732 feet.

Other major peaks: North Oscura Peak, 7,998 feet.

Dimensions: 4 x 27 miles.

Ecosystems: Chihuahuan Desert shrubs and grasses, including mesquite, creosote bush, yucca, ocotillo, prickly pear, and agave. At higher elevations, mountain mahogany, piñon-juniper, Gambel oak.

Counties: Lincoln, Socorro.

Administration: White Sands Missile Range, private.

Getting there: The only public access is to the BLM lands at the northern end of the range; call the BLM—Socorro Field Office for directions. Otherwise, access is restricted by White Sands Missile Range.

Non-Spanish speakers observing these mountains will want to mistranslate their name to "Obscure Mountains"—and for good reasons. Lacking the size and ecological significance of the San Andres Mountains, but like them placed mostly off-limits to the public by White Sands Missile Range, the Oscura Mountains probably would be ignored entirely were it not for one thing: In the shadow of the Oscura Mountains, on the morning of July 16, 1945, the world's first nuclear device was detonated, changing everything forever. Just 2.75 miles due east of Trinity Site, as chief scientist Robert Oppenheimer christened it, rise the increasingly steep slopes of the Oscura Mountains.

The name *Oscura* ("dark") actually derives from the Oscura Mountains' long and gentle eastern slope being more forested—and hence appearing darker—than the generally bare eastern escarpment of the San Andres Mountains. The Oscura Mountains are a large fault block—Precambrian rocks overlaid by thick sedimentary layers—tilting upward from the east, structurally akin to many ranges along the Rio Grande Rift. Thus, the eastern slopes are more gentle, have better soils, and retain water better than the steep western slopes.

Promising mineral leads were discovered in the Oscuras in 1876, with lead and copper the main ores. Northwest of Oscura Peak was the mining district of Hansonburg, organized around the camp there soon after 1900. At the same time, the little copper-mining camp of Estey City sprang up on the Oscuras' southeast side. Though commercial mining faded quickly in both areas and the camps were abandoned, rockhounds still make arrangements in Bingham to visit the privately owned Blanchard and Mex-Tex mines and collect samples. Specimens often are spectacular and include such rare minerals as atacamite, celestite, cerussite, cyanotrichite, linarite, murdochite, plattnerite, and spangolite. Common minerals include fluorite, calcite, barite, and quartz in collectible crystals.

Little Burro Mountains

Location: South of Oscura Gap, between Mockingbird Gap and the
Oscura Mountains.
Physiographic province: Basin and Range—Mexican Highland Section.
Elevation range: 5,200–6,400 feet.
High point: Unnamed summit, 6,505 feet.
Other major peaks: None named.
Dimensions: 2.5 x 8.5 miles.
Counties: Socorro.
Administration: White Sands Missile Range.
Getting there: There is no public access within the Missile Range.

These mountains—low hills, really—are a transitional group between the
Mockingbird Mountains, the northernmost of the San Andres Mountains,
and the Oscura Mountains. They are more closely allied with the Oscura
Mountains as evidenced by the Permian-Pennsylvanian rocks dipping eastward like the Oscuras, the western escarpment being the steepest.

San Andres Mountains

Location: South-central New Mexico, separating the Tularosa Basin from
the Jornada del Muerto.
Physiographic province: Basin and Range—Mexican Highland Section.
Elevation range: 4,200–8,000 feet.
High point: Salinas Peak, 8,965 feet.
Other major peaks: San Andres Peak, 8,236 feet; Silver Top Mountain,
8,140 feet, estimated; Big Gyp Mountain, 7,638 feet; Black Brushy
Mountain, 7,621 feet; Grandaddy Peak, 7,535 feet; Capitol Peak,
7,098 feet.
Dimensions: 11 x 85 miles, measured from San Augustín Pass to
Mockingbird Gap.
Ecosystems: Chihuahuan Desert shrubs and grasses, including mesquite,
creosote bush, yucca, ocotillo, prickly pear, and agave. At higher
elevations, mountain mahogany, piñon-juniper, Gambel and live
oak, and at the highest elevations, ponderosa pine.
Counties: Sierra, Doña Ana, Lincoln, Socorro.

Administration: White Sands Missile Range (WSMR).
Wilderness: 57,215-acre San Andres National Wildlife Refuge, extending
21 miles in the mountains' southern portion. No public access.
Getting there: You can't. The range, including its subranges, is entirely
within the WSMR.

In 1995, the wildlife biologists Kenneth A. Logan and his wife, Linda Sweanor, wrote at the conclusion of their landmark mountain lion research project in the San Andres Mountains:

> The establishment of White Sands Proving Ground (later White
> Sands Missile Range) in 1945 had the largest impact on the San
> Andres Mountains. It effectively ended the historic period of
> mining and ranching in the area, and any other forms of
> public use. After only 50 to 60 years of settlement, the San
> Andres Mountains were once again left essentially undisturbed
> by humans. Consequently, for the past 40 to 50 years, the
> mountains have been recovering naturally, making them,
> today, the largest single block of ecologically intact Chihuahuan
> Desert mountains remaining in southwestern North America.
> (*Desert Puma: Evolutionary Ecology and Conservation of an
> Enduring Carnivore,* 2001.)

Thus, thanks to the Cold War, New Mexico's fourth longest mountain range is completely insulated from the public, at least for now.

But human impacts can be subtle. In the mountains petroglyphs of the Mogollon Culture depict native deer and pronghorn, but they show no animals with the arching, saber-like horns of the oryx that now inhabit the mountains. Introduced into the White Sands Missile Range in 1969, the oryx have thrived. And while they prefer desert habitats to mountains, population pressures are pushing them into mountain canyons. In 1997, there were two thousand five hundred oryx on the WSMR. Barbary sheep, another exotic, also live on the WSMR. Javelinas were introduced into the mountains in the 1970s. In the 1990s, as many as one thousand eight hundred feral horses were removed from the missile range, though a few old stallions remain. Were the Mogollon Indians to return here, they would encounter a very different fauna.

Wildlife ecologists have feared that the exotics will expand at the expense of native species. Desert bighorn sheep are especially vulnerable. The San Andres Mountains are home to New Mexico's largest native herd of desert bighorn sheep. In 1976 they numbered two hundred animals. Then in 1979 they were scourged by scabies mites. By 1991 the disease had run its course, but as of 2002, the herd had only about thirty individuals. Late in 2002 fifty-one animals were transplanted to the range to bolster the population, with other transfers expected in the future. Most of the surviving desert bighorns are within the San Andres National Wildlife Refuge, established in 1941 specifically to protect desert bighorn habitat.

Native carnivores in the San Andres include coyotes, gray foxes, bobcats, badgers, striped skunks, ringtails—and mountain lions. During their 1985–95 study, Logan and Sweanor inventoried 100 lions living in the San Andres (81) and Organ Mountains (19).

The San Andres are a classic desert range. Only at the highest elevations are the mountains forested, and then sparsely. No perennial streams issue from the canyons—seeps and springs provide the most reliable water—and such streams able to escape the numerous rocky, normally dry canyons would disappear into the earth in the enclosed basins of the Jornada del Muerto on the west and the Tularosa Basin on the east. Fifteen thousand years ago, on cliffs overlooking the shores of a huge Pleistocene lake, Lake Otero, wood rats collected plant matter into middens that have survived to this day. The middens show that cooler and moister conditions then allowed a mixed conifer forest to grow at high elevations in the San Andres, but the middens also show that today's arid conditions have been in place at least four thousand years.

While the Sangre de Cristo Mountains of northern New Mexico are New Mexico's longest mountain range, the next three longest are all in south-central New Mexico: in descending order, Black Range, Sacramento Mountains, and San Andres Mountains. All were born with the tectonic upheavals at the beginning of the mid-Cenozoic development, 36 to 25 million years ago. Before that, layers of Precambrian, Lower Paleozoic, Pennsylvanian, Permian, and Cretaceous rocks were continuous across the region. Then Earth's crust stretched, the layers cracked along faults, and blocks rotated away from the basin, forming the graben (subsiding basin bounded by uplifting faults) known as the Tularosa Basin. At White Sands you can look west at the steep face of the San Andres Mountains, then turn around to face east and see the same formations on the steep face of the Sacramento Mountains.

The San Andres Mountains curve like a spine. Indeed, so regularly spaced are the east-west canyons that they rise to the crest like ribs to vertebrae. To the north, as the mountains become higher, they fragment into increasingly isolated units—Capitol Peak (7,098 feet), Fairview Mountain (6,890 feet), and the Mockingbird Mountains (see below). Farther north, across Mockingbird Gap, are the Oscura Mountains.

Several passes cross the San Andres Mountains: San Agustín Pass (5,720 feet), Rhodes Pass (6,528 feet), Mockingbird Gap (5,450 feet), Hays Gap (5,125 feet), Thurgood Canyon (5,430 feet), Lava Gap (4,590 feet), Sly Gap (5,450 feet), Hembrillo Pass (5,747 feet), Indian Trail Gap (6,764 feet), Woolf Gap (6,745 feet), and passes in Sulphur (5,917 feet), Dead Man (5,787 feet), San Andres (5,400 feet), Ash (5,803 feet), and Bear Canyons (6,370 feet).

Two subranges lie just south of Rhodes Pass. The Chalk Hills run eighteen miles north-south in a rough "S" shape and rise to almost eight thousand feet. The Hardscrabble Mountains, 3.5 miles east, extend 2.5 miles northeast-southwest and rise to six thousand nine hundred feet.

We know from Clovis Culture projectile points found north of the mountains that Mockingbird Gap was a major stop on the journeys of Paleo-Indians, just as it has been for travelers in the eleven thousand years since then. (Later travelers, however, didn't have to confront the Pleistocene lake that filled the Tularosa Basin then.) Beginning around eight thousand years ago, the Paleo-Indians were succeeded by hunter-gatherers of the Archaic Culture. By about 200 BC, at least some Indians were practicing sedentary living and agriculture. From AD 850–1350 the mountains were inhabited by people of the Jornada phase of the Mogollon Culture. Then, about AD 1400, these Puebloan peoples abandoned the San Andres Mountains, as well as all of southern New Mexico. We still don't understand why.

After a lacuna of about a hundred years, Apaches took possession of the mountains. They used the mountains to harvest game and wild plants. Later the mountains became a base for raiding, and then a refuge. During April 5–9, 1880, Apaches led by Victorio fought a major battle against the U.S. Army in Hembrillo Basin. The battle ended with the Apaches retreating to the Black Range, leaving behind breastworks, artifacts, and pictographs, some of people mounted on horses. The Apaches' presence also is recalled by the understated knob known as Victorio Peak (see below).

It was the Apaches—along with meager resources and harsh, arid terrain—that discouraged Hispanic settlement in the mountains. Don Diego de Vargas, in the journal he kept of his 1692 Reconquest, mentioned seeing the *sierra* of *Peñuelas,* "large rock without earth," and the perceptions of subsequent Spanish-speakers were no more sanguine. In the 1770s the mapmaker Bernardo Miera y Pacheco labeled the mountains *Las Petacas,* "the skin-covered chests," a plausible metaphor. No one seems to know how the mountains came to honor Saint Andrew, brother of Saint Peter and one of the twelve disciples. In the nineteenth and early twentieth centuries, the name appeared as both San Andres and *San Andreas,* but in 1920 the U.S. Board on Geographic Names approved a request by the Hon. Philip S. Smith to formalize the name San Andres.

In the 1880s and continuing to the 1930s, the mountains saw an influx of homesteaders and cattlemen. Most had small, hardscrabble operations, largely forgotten by history, but one produced the man with whom the San Andres Mountains always will be associated. In 1881 the family of the cowboy writer Eugene Manlove Rhodes moved from Nebraska to the mountains and settled at what became known as Rhodes Pass, on the main route between Engle and the Rio Grande on the west, and the Tularosa Basin on the east. Gene Rhodes taught school briefly in Alto (1890–91), but for most of his life here he was a working cowboy. He was a friend of many of the people who figured in the turmoil known as the Lincoln County War. When Oliver M. Lee and Jim Gilliland—friends of Billy the Kid—needed refuge after being charged with the murder of Col. Albert J. Fountain and his eight-year-old son, they turned to Rhodes, who later accompanied them to their eventual trial—and acquittal—in Hillsboro. The murders remain unsolved.

Rhodes drew upon many such experiences to write his stories, which were serialized by the *Saturday Evening Post* and earned him a national following. Sadly, this led him to spend the rest of his life at his wife's hometown in upstate New York, far from the country he so deeply loved. He returned only to die, in 1934. The headstone at his grave at Rhodes Pass, site of an annual pilgrimage allowed by the WSMR, reads *Paso Por Aqui*—"he passed by here"—the title of his most famous story.

The 1880s to the 1930s was also the period of mining in the San Andres, but as with homesteading and ranching, most operations were small, marginal. One geologist, Kingsley Charles Dunham, described mineralization as "of a decidedly feeble type." The main ores were galena, copper, barite, and quartz. Probably the greatest mineral contribution made by the mountains was the gypsum that dissolved from bedrock formations, later reconstituted in the evaporite lakebeds of the central Tularosa Basin. The gypsum-laden runoff waters flowed into the Pleistocene lake known as Lake Otero. With warmer, drier times, the lake shrank to small, shallow remnant lakes, including the one known as Lake Lucero. Then prevailing southwesterly winds deposited gypsum fragments picked up from the dried beds of Lake Lucero—creating White Sands.

But mining and ranching and all other public activities ended abruptly in 1945 when WSMR was established. As Logan summarized, "After only 50 to 60 years of settlement, the San Andres Mountains were once again left essentially undisturbed by humans." Nothing in the foreseeable future seems likely to alter that.

San Agustín Mountains

Location: Immediately north of San Augustín Pass on US 70, two miles
    east of the village of Organ.
Physiographic province: Basin and Range—Mexican Highland Section.
Elevation range: 5,200–6,700 feet.
High point: San Agustín Peak, 7,010 feet.
Other major peaks: None named.
Dimensions: 3 x 4.5 miles.
Ecosystems: Lower Sonoran grasses and shrubs, including mesquite.
Counties: Doña Ana.
Administration: White Sands Missile Range.
Getting there: No public access within the Missile Range.

Like the Mockingbird and Hardscrabble Mountains, these are a small subrange within the much larger San Andres and would attract little attention were it not for the dramatic presence of San Agustín Peak overlooking US 70.

Victorio Peak

Location: In the western foothills of the San Andres Mountains, just
    south of the Sierra-Doña Ana County line.
Physiographic province: Basin and Range—Mexican Highland Section.
Elevation: 5,525 feet (estimated).

Relief: 225 feet.

Counties: Doña Ana

Administration: White Sands Missile Range.

Getting there: You can't. Not without permission from military
    officials at White Sands Missile Range, and they aren't eager to
    hear requests about getting to Victorio Peak.

Just as New Mexico's most well-known contemporary incident—the Roswell
UFO crash—is the source of controversy as to whether it ever happened, so
the tales behind one of New Mexico's most talked about mountains also are
cloaked in skepticism and controversy.

Victorio Peak, a modest, 5,525-foot, isolated prominence in the
Hembrillo Basin on White Sands Missile Range, is an unlikely candidate for
legend, but it suited Milton E. "Doc" Noss when he returned from a hunting trip there in 1937 claiming to have found a fabulous cache of treasure.
(According to an El Paso man, Noss had earlier approached him about a
scheme to salt a mine in the Caballo Mountains to attract wealthy investors.
The man refused.) According to Noss, a passage hidden under a rock at the
mountain's top led to a chamber a train could drive through, filled with
everything from Spanish gold to religious artifacts to Spanish documents
to Wells Fargo chests—gold bars stacked like cordwood. Originally valued
at $26 million, the trove is valued at $3 billion by today's treasure hunters.
Guarding the treasure were twenty-seven skeletons; proponents now say
seventy-nine skeletons are there. (They've been reproducing.)

Noss showed the site to his wife and supposedly brought out several
hundred gold bars, along with jewels. But frustrated by a constriction in
the passage, Noss had it dynamited to enlarge it. Instead, the blast sealed
the treasure chamber. The succeeding decades have been a tortuous succession of efforts to re-enter the chamber, involving mining engineers,
investors, and many others. In 1949, a disappointed investor accused Noss
of fraud. In the following altercation, Noss was shot dead. The shooter
pleaded self-defense and was acquitted.

Nonetheless, the legend grows, and the treasure recovery efforts continue. In the 1970s, the flamboyant attorney F. Lee Bailey successfully sued
White Sands Missile Range on the behalf of Noss family members for permission to probe for the treasure. Despite disappointing results, they keep trying, though White Sands Missile Range is reluctant to grant permission again.

Southwestern New Mexico—Basin and Range

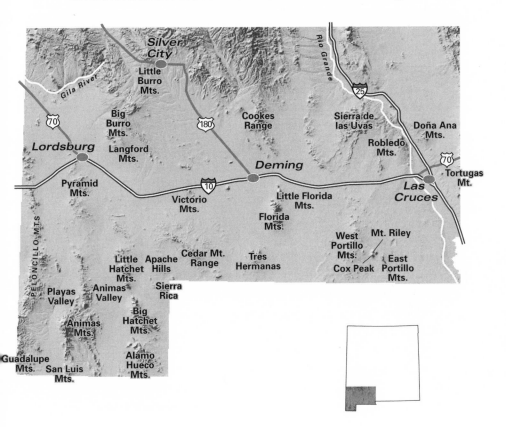

Tres Hermanos

Location: In the Tularosa Basin southwest of Alamogordo.

Physiographic province: Basin and Range—Mexican Highland Section.

Elevation: Twin Buttes, 4,553 and 4,408 feet; Lone Butte, 4,358 feet.

Relief: 350–550 feet.

Ecosystems: Lower Sonoran shrubs and grasses, especially mesquite.

Counties: Otero.

Administration: BLM—Las Cruces Field Office, state, private.

Getting there: Check with the BLM regarding the current status of roads
and gates; respect private property.

The *Tres Hermanos,* "three brothers," are a constellation of three small
knobs, each rising abruptly from the otherwise featureless Tularosa Basin.
Unlike other sibling formations in New Mexico, these are not contiguous.
Twin Buttes, barely 0.25 mile apart, are 3.5 miles west of Lone Butte. Twin
Buttes consist of Permian limestone and, at the north brother, sandstone
formations. Lone Butte is a mid-Cenozoic igneous intrusion, monzonite
intruding grandiorite. The three erosional remnants are part of the larger,
east-dipping structural block of the Jarilla Mountains to the south and
Tularosa Peak to the north in the Tularosa Basin.

Jarilla Mountains

Location: Southwest of Alamogordo, west of US 54, north-northwest of
Orogrande.

Physiographic province: Basin and Range—Mexican Highland Section.

Elevation range: 4,260–5,200 feet.

High point: Unnamed summit, 5,301 feet.

Other major peaks: None named.

Dimensions: 6 x 10 miles.

Ecosystems: Chihuahuan Desert scrub, including creosote bush,
mesquite, yucca, and cacti.

Counties: Otero.

Administration: BLM—Las Cruces Field Office, state, private.

Getting there: From Orogrande several dirt roads lead to mines in the
Jarilla Mountains.

The Jarilla Mountains are among the few New Mexico mountains where Indians mined turquoise before Europeans arrived. In 1879, American prospectors found not only turquoise but also iron, copper, gold, and silver in the mountains' monzonite porphyry and carboniferous limestone formations. Serious mining, however, waited until early 1890, when the Southern Pacific RR branch to Carrizozo passed by the Jarilla Mountains' eastern flank. The gold rush of 1906–14 attracted two thousand people to the mountains and the nearby town, which changed its name from Jarilla Junction to Orogrande, "big, or plenty, gold." Herbert Ungnade reported that iron ores of hematite and magnetite were shipped to Pueblo, Colorado, from 1916 to 1921, and that gold, silver, copper, and lead ores mined from 1904 to 1929 were valued at $12 million. Today, the mines are moribund, though they continue to attract mineral collectors.

The mountains and their mineralization resulted when an igneous mass rose toward the surface in mid-Cenozoic time, shoving upward the overlying Paleozoic sedimentary layers. This created an elongate north-south dome, with canyons and ridges radiating outward. The strata include Pennsylvanian rocks near the south end of the mountains.

Monte Carlo Gap, a low pass (4,232 feet in elevation), divides the mountains into two sections. The smaller northern section reaches 5,295 feet; the larger southern section has higher elevations. The name *Jarilla* is derived from the Spanish *jara,* "willow." The hydrophillic willow is a curious eponym here, as there's little or no water in the mountains, and the dominant vegetation is cactus, yucca, mesquite, creosote bush, and grasses. During the mining era, wooden pipes were laid across the desert basin to the east to bring water to the Jarilla Mountains from the Sacramento Mountains, thirty-six miles away.

Fra Cristobal Range

Location: East of the Rio Grande, northeast of Elephant Butte
　　Reservoir, bounded on the east by the Jornada del Muerto,
　　south of the Sierra-Socorro County line, north of NM 52.
Physiographic province: Basin and Range—Mexican Highland Section.
Elevation range: 4,800–6,600 feet.
High point: 6,834 feet.
Dimensions: 3 x 16 miles.

Other major peaks: Fra Cristobal Mountain, 6,003 feet.

Ecosystems: Mostly Chihuahuan grasses and shrubs, including grama
and bluestem, creosote bush, ocotillo, and mountain mahogany,
occasional piñon-juniper at higher elevations.

Counties: Sierra, Socorro (barely).

Administration: Private (part of the Pedro Armendaris Ranch, owned by
Ted Turner's New Mexico Ranch Properties).

Each day hundreds of people driving on I-25 pass the Fra Cristobal Mountains, and a few, perhaps, glance at the complex, rugged western escarpment
of these desert mountains—remote, forbidding, torturous with canyons and
ridges—and wonder just what tales the mountains could tell, what treasures
are hidden in the mountains' caves.

Well, some of the mountains' tales are known, such as that of the epony-
mous Franciscan father, Cristóbal de Salazar. He was Don Juan de Oñate's
cousin and served as *sargento mayor* on Oñate's colonizing expedition of
1598. Legend says that Fra Cristóbal died near these mountains while returning to Mexico, leaving his name on the place of his demise. Something similar happened near the Robledo Mountains (see entry), and Don Diego de
Vargas recorded in his campaign journal of 1692 that from *El Paraje de Fra
Cristóbal* on *El Camino Real* he could see the mountains called *El Muerto*,
"the dead man," identified since as the Fra Cristobals.

As for the caves, their treasure has been found: bat dung. The Jornada
bat caves in lava tubes near the Fra Cristóbals host one of North America's
largest bat colonies, with more than ten species represented and, in heavy
insect years or during migrations, as many as 5 to 8 million individual bats.
Bat dung, or guano, was mined from limestone solution caves during the
twentieth century.

The Fra Cristóbals have other ecological treasures besides bats. The
mountains are entirely within the sprawling land grant given in 1820 to
Pedro de Armendaris, a military officer stationed in Santa Fe. He abandoned the land when he left New Mexico for Chihuahua. For this and other
reasons, the mountains escaped heavy cattle and sheep grazing. This was
reinforced in 1993 when the media magnate Ted Turner purchased the land
grant and, as with his holdings elsewhere in New Mexico, directed that the
lands be managed to emphasize natural values and native species. On the
mountains' high ridges flourishes New Mexico feather grass, a cool-weather

species all but gone from most of southern New Mexico. Thus, the mountains offered near-pristine habitat for the thirty-five desert bighorn sheep reintroduced to the mountains in 1995; by 1998 the herd had grown to forty-five animals.

Geologically, the Fra Cristobals were formed, like many other Rio Grande Rift mountains, by faulting and uplift of a block of sedimentary layers. The mountains also have a history of thrust uplifting from the time of the Laramide Orogeny 70 million years ago. Most of the layers exposed on the western escarpment are Paleozoic, lying atop Precambrian granite. The Fra Cristobal uplift is rather small, but it's well defined by faults, appearing on a map like a laurel leaf, tapering to points at its northern and southern ends. Because of the mountains' long history of private ownership, hiking in the mountains has been minimal, or less. So for the I-25 drivers who wonder about the mountains, they should stop and get a good look, because that's likely to be as close as they'll get.

Caballo Mountains

Location: Running generally south from Truth or Consequences to
　　Rincon, east of the Rio Grande.
Physiographic province: Basin and Range—Mexican Highland Section.
Elevation range: 5,000–8,300 feet.
High point: Timber Mountain, 7,565 feet.
Other major peaks: Brushy Mountain, 7,375 feet; Caballo Cone, 6,081 feet.
Dimensions: 10 x 28 miles.
Ecosystems: Desert scrub, including creosote bush, mesquite, and
　　mountain mahogany, with desert grasses and then juniper as
　　elevation increases.
Counties: Sierra, Doña Ana.
Administration: Mostly BLM—Las Cruces Field Office, state, private.
Wilderness: Caballo Mountains Wilderness Complex, 79,616 acres, on the
　　range's southeast side, proposed by the NM Wilderness Alliance.
Getting there: Access generally is poor; east and west slopes are reached,
　　when at all, by a maze of unmarked dirt roads, often very rough.
　　To reach the range's northwest side from T or C, take NM 51 across
　　the Rio Grande, then turn right. To reach the mountains' east slopes,
　　take NM 51 east to Engle, then head south over the all-weather,

graded road that parallels the mountains along the Jornada del Muerto. From this road, rougher dirt roads head west toward the mountains, including one to the summit of Timber Mountain. The road along the Jornada ends on the south at I-25 Exit 32, just east of Rincon. To reach the mountains from the south and west, drive across the dam of Caballo Reservoir to numerous dirt roads in the mountains' western foothills.

The Caballos are the sister range of the Fra Cristobal Range immediately north, and first-cousin to other fault-block mountains along the Rio Grande Rift. All present a characteristic geologic configuration: steep escarpments facing the rift, gentler slopes away from it. The uplift that created the Caballo Mountains began in the late Oligocene-early Miocene, about 24 million years ago. In the Caballos, the western escarpment rises three thousand three hundred feet above the river in a seemingly vertical gallery of cliffs. On their faces the uplift has exposed ancient Precambrian rocks, overlain by only-slightly-less-ancient Paleozoic layers, mostly limestone. On the Caballos' gentler eastern slopes, Paleozoic limestone layers outcrop in long, sinuous lines snaking through the stippled vegetation. The uplift pattern emerged when the rift began to spread; the center subsided, and adjacent blocks were tilted upward along the Caballo Fault, visible at the base of the mountains' west face. Geologists say the overturned folds and complex faults in the Caballos and Fra Cristobals are apparent even to non-geologists.

This geological complexity accompanies ecological complexity. The Caballo Mountains are within the Lower Sonoran and Upper Sonoran life zones; common Chihuahuan Desert plants are found here including mesquite, ocotillo, creosote bush, sotol, soaptree and other yuccas, and numerous cacti. The "state-sensitive" plant, *Penstemon ramosus,* has been reported here.

In 1996, the NM Department of Game and Fish found the Caballo Mountains to be suitable habitat for reintroduction of desert bighorn sheep, but opposition from citizens in T or C scuttled the plan. (The sheep were reintroduced to the Fra Cristobal Mountains just to the north.)

The Caballo Mountains form the western boundary of the *Jornada del Muerto,* "journey of the dead man" (not "journey of death," as often mistranslated). On the west, the range overlooks the Rio Grande and its flanking deep canyons and steep-sided, sand-filled arroyos. It was these deep arroyos,

Palomas Gap and the old toll road through the Caballo Mountains

along with dense vegetation and mud along the Rio Grande, that motivated *carreteros* on El Camino Real to risk taking their cumbersome wagons to the dry-but-flat Jornada to the east instead.

Approximately midway along their length, the Caballos are divided by the deep cleft known as Palomas Gap. South of the gap, the mountains attain their highest point, 7,565 feet at Timber Mountain; the lower northern section culminates in Caballo Cone, 6,081 feet in elevation, atop Turtleback Mountain. Palomas Gap itself is an impressive feature. Around 1900 a toll road was constructed through here to connect the Rio Grande Valley and the Black Range mining camps with communities to the east, such as White Oaks. The collapse of mining, the creation of Caballo Reservoir in 1937–38, and the establishment of White Sands Missile Range after World War II all doomed the road, though its stone-buttressed bed still runs high on the gap's steep south side.

Around 1901 the Caballos caused a stir statewide when gold was discovered in gravels at the range's southwestern end, around Shandon and Trujillo gulches. Though the excitement faded when the few gold-bearing veins soon played out, amateur prospectors can still get "color" with dry panning. (Not surprisingly, the Caballos have their lost treasure legend. It tells of Indian raiders stashing their loot—a trove of gold bars—beneath rocks near a spring in Caballo Canyon. Seeking the treasure is complicated by the Caballo Mountains having two Caballo Canyons.)

In 1909 the mountains again attracted attention when the nation's only workable deposit of vanadium was mined here, though the veins proved shallow. Other minerals mined in the Caballos have been lead, fluorspar, and manganese. Today, weathered buildings, heaps of rubble, abandoned mines, and feral roads are all that remain.

Few hikers explore the Caballos. Access is poor, the terrain is tough, water is lacking, and the mountains are hellishly hot in summer. Furthermore, most of the roads here are very rough—and all are unmarked. A person would find a GPS—or very good directions—useful in following a route in this maze. Trails also are lacking; supposedly one exists ascending Turtleback Mountain, but it's difficult to find. For most destinations trails aren't really necessary, as the area is laced with jeep tracks and old mine roads. Cross-country hiking also is straightforward, over a generally "open" landscape with landmarks readily visible. The terrain is surprisingly scenic and interesting—the flora of the Chihuahuan Desert, abundant and diverse

Outstanding in his field: the author and a sotol plant in the western Caballo Mountains

bird life, desert mammals and reptiles, and the decaying remains of mining. (*Never* enter an abandoned mine; they're unspeakably dangerous!)
And for those seeking solitude, there's little likelihood of encountering
other hikers here. On the range's southeast side is an approximately eighty
thousand-acre area that has been proposed for wilderness by the NM
Wilderness Alliance.

The range's name, *Caballo,* "horse," likely originated from wild horses
in the mountains, though a formation at the range's northern end supposedly resembles a horse's head (my efforts at seeing it have failed). In 1766 the
range was recorded as *El Perrillo Sierra,* "the little dog range," likely for *Ojo
del Perrillo* on the range's east side. But in 1807 Zebulon Pike mentioned
passing the west side of "Horse Mountain," and an 1828 map labeled the
range *Sierra de los Caballos.* When you view the Caballo Mountains from
the north, whether from I-25 or from T or C, look for Turtleback Mountain.
It doesn't take much imagination to see a turtle sprawled atop the highest
point at the range's north end.

Doña Ana Mountains

Location: 8 miles north of Las Cruces, east of I-25.
Physiographic province: Basin and Range—Mexican Highland Section.
Elevation range: 4,500–5,800 feet.
High point: Doña Ana Peak, 5,829 feet.
Other major peaks: Summerford Mountain, 5,754 feet.
Dimensions: 4 x 7 miles.
Ecosystems: Chihuahuan Desert shrubs and grasses, including mountain
    mahogany, Emory oak, manzanita, mesquite, creosote bush, and
    scattered juniper.
Counties: Doña Ana.
Administration: BLM—Las Cruces Field Office, NM State University
    (for ecological research).
Getting there: From Exit 6 on I-25 head east on US 70 for 3.3 miles to where
    Jornada Road heads north. This becomes good dirt. At
    6.6 miles from US 70 you reach the boundary of the USDA Jornada
    Experimental Range. Here, immediately south of the boundary, a
    rougher but still passable road heads west and after about 3 miles
    arrives in the basin at the heart of the Doña Ana Mountains.

The impression the Doña Ana Mountains make greatly surpasses their size. A small knot of mountains, with relief of only one thousand three hundred feet, parched and harsh, the mountains can be as forbidding and compelling as a lithograph by Gustave Doré illustrating Dante's *Inferno*. They are a classic desert range—a barren, exposed, oven-red mass of rock eroded into shapes born in the realms of fantasy.

Appropriate, then, that the mountains are the remnants of what was once a fiery ring, a geological structure called a *caldera*. During the mid-Cenozoic magma chambers began pushing toward the surface here, cracking and uplifting the overlying sediments. From south of Las Cruces north to the Caballo Mountains, a great block of sedimentary rocks east of the Rio Grande began tilting upward as part of the expanding Rio Grande Rift. As this was happening, two volcanoes erupted at the site of the present Doña Ana Mountains, spewing out the rhyolite and ash-flow tuffs whose eroded remains are the sculpted formations seen today. Their two calderas—the Dagger Flat caldera and the much-larger Doña Ana caldera (35.5 million years old)—filled with debris during subsequent millennia, yet still today a person standing in the roughly circular valley at the mountains' center can feel close to the mountains' earlier history.

On the west, many small canyons cut into the mountains. On the east and north canyons are fewer and broader. The range consists of two main groups: in the south one associated with Doña Ana Peak, and in the north one associated with Summerford Mountain, a laccolith related to the Dagger Flat caldera. Though early prospectors reported finding small but very rich pockets of gold-silver ore, the reports either were false or greatly exaggerated, or the pockets were soon exhausted.

No developed hiking trails exist in the Doña Ana Mountains, but there are numerous dirt roads, and cross-country hiking is fairly easy across the open terrain.

Organ Mountains

Location: East of the Rio Grande, east of Las Cruces, south of US 70.
Physiographic province: Basin and Range—Mexican Highland
   Section.
Elevation range: 4,600–9,000 feet.
High point: Organ Needle, 9,012 feet.

Other major peaks: Organ Peak, 8,870 feet; Rabbit Ears, 8,150 feet;
 Baylor Peak, 7,721 feet.

Dimensions: 8 x 17 miles.

Ecosystems: Chihuahuan Desert shrubs, including agave, ocotillo,
 mesquite, mountain mahogany, sotol, alligator juniper, piñon,
 and at higher elevations ponderosa pine, and finally, isolated Douglas
 fir and white fir.

Counties: Doña Ana.

Administration: BLM—Las Cruces Field Office, White Sands Missile
 Range.

Wilderness: Organ Mountains BLM Wilderness Study Area, 7,283 acres.

Getting there: Access from the west is best from the BLM A.B. Cox Visitor
 Center, reached by going east on University Blvd. at the south end
 of Las Cruces. This becomes the pavement and good dirt Dripping
 Springs Road, for 10.5 miles. Access from the east is via US 70, just
 east of San Agustín Pass, where a marked, paved road leads 5.5 miles
 to the BLM Aguirre Springs National Recreation Area.

At the Organ Mountains' western base, near the BLM A.B. Cox Visitor Center,
was an old, weathered wagon, and nearby a Spanish bayonet yucca, complete with flower stalk. With the Organ Mountains as background, the effect
was stunning, and with a little wiggling a photographer could usually fit all
three—wagon, yucca, and mountains—into the frame. Through countless
photos like this the Organ Mountains have come to symbolize New Mexico's
desert ranges—and deservedly so, for few desert ranges approach the drama
of the Organs' spires, peaks, and domes. And unlike many desert ranges
along the Rio Grande, the vertical spectacle is undiminished whether viewed
from the east or the west. (The Tularosa Basin at one thousand feet lower
than the Rio Grande Valley accentuates the eastern relief.)

The Organs are part of an uplift of a 150-mile, west-tilted fault block that
extends from the Mexican border into central New Mexico. In the Organs,
Precambrian granite overlain by as much as eight thousand five hundred
feet of Paleozoic and Cretaceous strata was thrust upward during the
Laramide Orogeny, 70 million years ago, and shed sediments into surrounding lowlands.

Then about 36 million years ago, during mid-Cenozoic time, the upper
part of a magma chamber intruded into these strata and erupted. These

Organ Mountains, west side

andesitic volcanoes collapsed, forming the Organ Caldera (35.8 million years ago) and underlying batholith. Then all this was rotated on its side. The Paleozoic strata can still be seen on the range's western base and along Rattlesnake Ridge in the southeast. But it was the vertically jointed granite, dominating the range's central portion that gave the range its distinct character—and its name.

The Spanish explorer-colonizer Don Juan de Oñate, who passed by here in 1598, noted the mountains and with wry humor called them *Sierra de Olvido,* "mountains of forgetfulness," because members of his party, who'd seen them before, could offer no recollections of them. But when Governor Otermín passed by in 1682, he referred to them as *Los Organos,* for their resemblance to organ pipes. Other early Spanish names included *La Sierra Grande de los Mansos,* "the big range of the fingers," and *La Sierra de Soledad,* "the mountains of solitude."

With important water sources at Dripping Springs on the west and Aguirre Springs on the east, the Organs would have been well known to early Indians and also to early Hispanic settlers, but the Indians left few

Rabbit Ears, Organ Mountains

traces, and the Hispanics stayed near the Rio Grande. That is unless you give credence to the legend of the Lost Mine of Padre La Rue (see sidebar).

More verifiable mining began in 1849, with the discovery of the lead-zinc ore body that led to the Stephenson-Bennett Mine. Mining intensified around 1900, then waned, and despite a brief flurry of activity at the beginning of World War II, was soon dead. Yet up to 1942, the Organ Mining District produced more than $2.5 million (1936 value) from copper, lead, silver, zinc, and gold, in descending order. More than half of that came from the Stephenson-Bennett Mine. Mines in the Bishop Cap area produced more than five hundred tons of fluorite. While silver and other metals had been the objective when the Stephenson-Bennett Mine was opened, miners soon discovered a large chamber with spectacular deposits of calcite, calcite stalactites, and aragonite, and crystals of wulfenite and cerussite were of such quality that the mine made revenue as mineral collectors worldwide bid for the specimens.

Today, outdoor recreation and ecosystem preservation are the main activities in the Organs. The primary trailheads are at the BLM A.B. Cox

Pitch 7 on Sugarloaf Peak, Organ Mountains

The Lost Mine of Padre La Rue

Legend tells that sometime before 1800, at a hacienda in Chihuahua, a Franciscan priest named La Rue received from a dying friend details about rich gold deposits two days northward. When famine struck his parishioners, Father La Rue searched in the Organ Mountains until he recognized the landmarks that led him to the gold. He brought his flock north to settle near Spirit Springs, now the BLM A.B. Cox Visitor Center. Tradition holds that from there they worked a mine located in a deep canyon to the east or southeast of the springs.

But the Church, angry at La Rue's decampment and covetous of the gold, sent soldiers. La Rue and his followers hid the mine, and despite torture and the murder of La Rue they refused to divulge its location, thus adding it to the treasure horde of lost mine stories in the Southwest. Colonel A. J. Fountain of Las Cruces was among the seekers, and shortly before his murder, still unsolved, he claimed to have found it. This has only spurred the treasure hunters, who have not been deterred by lack of evidence of a Padre La Rue in Chihuahua nor by geological surveys and extensive prospecting revealing no significant traces of gold.

Visitor Center on the west and the BLM Aguirre Springs National Recreation Area on the east. The 5.8-mile Baylor Pass National Recreation Trail, connecting east and west sides, also is popular. Before it was called Baylor Pass, commemorating Confederate Army Colonel John R. Baylor, who used the pass to surprise and capture about seven hundred Union troops in 1861, the pass was known as the Old Salt Trail, because livestock were driven east over it to the salt flats between El Paso and Carlsbad.

The range's small size, constricted even further by being partly within White Sands Missle Range on the east, discourages backpacking, but day hikes are spectacular. Among rock-climbers, the Organs are famous, not only for the scenery but also for the excellent quality of the rock and the abundance and variety of climbing routes. Some summits in the Organs, such as Organ Needle, Rabbit Ears, and Sugarloaf, are reached only by technical climbers. Sugarloaf can require as many as fourteen pitches for the ascent.

The foothills are a showcase of Lower Sonoran life zone plants and shrubs: agaves, sotol, desert willow, acacia, netleaf hackberry, creosote bush, ocotillo, and yucca. Within a short distance, Upper Sonoran life zone plants such as alligator juniper, piñon pine, Apache plume, and mountain mahogany appear. Still higher, in the Transition Zone, are ponderosa pines, and in the highest, coolest sites are a few representatives of the Canadian Zone such as Douglas fir and white fir. The mountains are home to several species of endangered cacti, including the Organ Mountains pincushion cactus. In all, twelve plant species, including the rare Organ Mountain evening primrose, are found only in these mountains.

Animals are similarly diverse and include some 80 species of mammals, 185 species of birds, and 60 species of reptiles and amphibians. The Organ Mountain woodland snail is among several rare species. An isolated race of the Colorado chipmunk lives here. Montezuma quail and mule deer are common, and ringtails often frequent campsites at night.

Extremely difficult terrain and awkward access in the south and east have left significant portions of the Organs wild, despite the range's small size and proximity to Las Cruces. But Las Cruces now is the second largest city in the state, and growing explosively. The time of the Organs being "forgettable mountains" has passed.

Tortugas Mountain

Location: Just east of I-25, at the south end of Las Cruces.
Physiographic province: Basin and Range—Mexican Highland Section.
Elevation: 4,931 feet.
Relief: 700 feet.
Ecosystems: Lower Sonoran shrubs and grasses, including mesquite, creosote bush, ocotillo, and yucca.
Counties: Doña Ana.
Administration: New Mexico State University.
Getting there: From the southeast section of Las Cruces, University Blvd. heads east toward the mountain with an "A" on it. The paved highway reaches the mountain's northern end at 2.25 miles.

This is one of those mountains whose significance is independent of its height or size. History has passed within the shadow of this formation near the Rio Grande and left its mark—sometimes literally. Many people have attributed its Spanish name, meaning "turtles," to a nearby Indian village of the same name. Each December villagers make pilgrimages to the bonfire-lit summit to honor the Virgin of Guadalupe. Other people say the rounded limestone formation resembles a turtle's shell (see *Turtleback Mountain* in the Caballo Mountains), especially when viewed from the south. Perhaps the name was derived from an incident involving turtles, or perhaps it is a translation of an Indian name.

The formation's other name, A Mountain, is easy to explain. Students at the state's agricultural school, now New Mexico State University, in a rite almost obligatory at colleges everywhere, painted a huge "A" on the mountain's side, where it remains today (see *M Mountain* in the Socorro Mountains). NMSU continues to administer a thicket of antennas, electronic transmitters, and observatories on the summit. A complex network of steep, rough trails is used by hikers, mountain-bikers, equestrians, four-wheelers, and others.

Geologically, Tortugas Mountain is an uplifted and faulted block of Permian limestone, tilted to the southwest. Early in the twentieth century, prospectors found fluorite veins on the mountain. They sunk shafts as deep as 150 feet, and an aerial tram took ore down the mountain. The mines had played out by 1935, but the mountain's value as a school emblem is inexhaustible.

Franklin Mountains and Bishop Cap

Location: East of the Rio Grande, paralleling I-10 from Anthony to El Paso.

Physiographic province: Basin and Range—Mexican Highland Section.

Elevation range: 4,000–7,000 feet; New Mexico portion, 4,200–5,100 feet.

High point: North Franklin Mountain, 7,192 feet; New Mexico portion, North Anthonys Nose, 5,388 feet.

Other major peaks: Anthonys Nose, 6,917 feet; Bishop Cap, 5,419 feet.

Dimensions: 3.5 x 23 miles.

Ecosystems: Chihuahuan Desert shrubs and grasses, including creosote bush, mesquite, ocotillo, cacti, yuccas.

Counties: Doña Ana.

Administration: BLM—Las Cruces Field Office; White Sands Missile Range.

Getting there: The New Mexico portion of the Franklin Mountains is accessible from Anthony Gap, reached by driving south from

Las Cruces to Exit 162, the first Anthony exit. From here NM 404 heads east 5.25 miles to the gap, elevation 4,424 feet. In Texas, the Transmountain Highway goes between South Franklin Mountain and Franklin Mountain to connect the eastern and western prongs of the city.

New Mexico has but the northern tail of this range, yet its crest dominates the eastern skyline as one travels I-10 south from Las Cruces to El Paso. It's a jagged, narrow ridge beneath which writhe, sinuous limestone layers—a textbook illustration of sedimentary layering. (Actually, the layers writhe less than they seem, an optical illusion of the sculpted slopes.)

Less than one-third of the narrow linear range's twenty-three-mile length are in New Mexico. Political geography decreed that the rest, including the highest point, would be in Texas. Perhaps that's appropriate, as the range most likely takes its name from an early name for El Paso—Franklin—named for Franklin Coons, an early American resident. The New Mexico section extends about 7.5 miles north from Anthony Gap to Fillmore Pass. Long ago, the Rio Grande spilled through the pass into the Tularosa Basin. About 5.5 miles farther north are the outliers Bishop Cap, 5,419 feet, and Pyramid Peak, 5,261 feet. Bishop Cap is said to resemble a bishop's mitre, and indeed it does, especially when viewed from the northwest. (Anthonys Nose farther south is said to resemble the saint's nose; any resemblance is hardly obvious.) These outlier peaks are located near the Organ Mountains, but geologically they're part of the Franklins. They are made up of west-tilted Paleozoic sedimentary layers, mostly fossiliferous limestone but also shale and siltstone. The Chihuahuan Desert vegetation here includes creosote bush, mariola, lechuguilla, ocotillo, yucca, and cacti. Bishop Cap has an especially rich assemblage of cacti, while around North Anthonys Nose are lechuguila agaves, signature plants of the Chihuahuan Desert.

No developed trails exist in the Franklin Mountains, and there are few hikers. To reach the top of Bishop Cap, leave I-10 at Exit 131 for Mesquite. Drive beneath the highway, then jog right to reach a dirt road heading into Mossman Arroyo. After 4.4 miles, the road forks. Park here and ascend the ridge immediately to the south, climbing steeply over Pyramid Peak to reach Bishop Cap after two miles and two thousand feet. The point of North Anthonys Nose is reached from Anthony Gap by hiking north along the mountain's west base for 1.2 miles before ascending a ridge and following it 2.1 miles to the top.

Sierra de Cristo Rey

Location: West of El Paso, just north of the Mexican border.
Physiographic province: Basin and Range—Mexican Highland Section.
Elevation: 4,567 feet.
Relief: 825 feet.
Ecosystems: Chihuahuan Desert scrub.
Counties: Doña Ana.
Administration: Private.
Getting there: Sierra de Cristo Rey is reached by taking NM 27 east
    from Sunland Park on the south side of the Rio Grande.

This small, singular peak, located on what appears to be an empty lot in the neighborhoods of both El Paso and Ciudad Juarez, seems to belong more to Mexico and the Sierra Juarez than to New Mexico, and it would claim little attention were it not for the 45-foot-tall statue of Christ on its summit. This was placed there in 1935, at which time the peak's name changed from *Cerro de los Muleros,* "hill of the mule drivers," to "mountains of Christ the King." A series of rocky switchbacks takes pilgrims to the summit shrine over a well-worn path.

The peak has more than religious significance. Geologically, it is a Cenozoic intrusion into older Cretaceous sediments, and erosion has exposed the tracks of dinosaurs.

Sierra Ladrones

Location: 45 air miles south of Albuquerque, west of the Rio Grande,
    west of Bernardo.
Physiographic province: Basin and Range—Mexican Highland
    Section.
Elevation range: 5,500–9,000 feet.
High point: The twin summits of Ladron Peak, 9,210 feet westernmost,
    and 9,143 feet, eastern.
Other major peaks: Monte Negro, 7,581 feet; Cerro Colorado, 5,736 feet.
Dimensions: 8 x 13 miles.
Ecosystems: Piñon-juniper, mountain mahogany, creosote bush,
    high-desert grasses and shrubs.
Counties: Socorro.

Administration: BLM—Socorro Field Office, Sevilleta National Wildlife
   Refuge, private.
Wilderness: Sierra Ladrones BLM Wilderness Study Area, 45,308 acres.
Getting there: From Exit 175 on I-25 at Bernardo, a paved road heads
   west across the Rio Puerco, soon becoming dirt County Road B12.
   This loops north around the mountains and eventually arrives at
   the ghost town of Riley, far to the southwest on the Rio Salado.
   From this county road several dirt roads head toward the
   mountains, but none is marked as an access, and a BLM or USGS
   map is essential. These roads likely terminate at private property,
   but until then they do get you a little closer to the mountains.
   Also, some roads have been closed following introduction of
   desert bighorn sheep to the mountains.

Perhaps no mountains are so immediate to a large urban area yet so remote
as the Sierra Ladrones are to Albuquerque. From neighborhoods throughout the city, from Interstates 25 and 40, the Ladrones are visible to the south,
their jagged summits sometimes seeming to float atop the valley haze. Yet
while it's just an hour's drive to the mountains' foothills, all familiarity stops
as soon as pavement becomes dirt. The Ladrones host no campgrounds, no
trailheads, no trails, certainly no roads—no human habitations. From any
ridge or summit in the Ladrones, Albuquerque is but a small, hazy smudge
far in the distance.

   I once wrote a story for the *Albuquerque Journal* about the Ladrones.
Later a woman called to tell me that as a youth she'd been a cook at a line
camp on her father's cattle allotment, which included the mountains. When
I described my little hike to her, she exclaimed, "You didn't go into those
mountains alone, did you? Oh, you must never go into the Ladrones alone."
Then she told stories from her childhood to convince me. But it wasn't hard;
the Sierra Ladrones just naturally have a menacing presence: brutally rugged,
confusingly complex, arid, and salted with dark and glittering legends.

   The name *Sierra Ladrones* is Spanish for "thieves' mountains." In the
days when Navajos and Apaches raided villages along the Rio Grande, they
sometimes drove stolen stock into these mountains, where among the tortuous and difficult canyons the "thieves" were safe from capture. As *The
Place Names of New Mexico* explains, "Later, non-Indian rustlers and high-
waymen used these rugged mountains as a hideout, and legends abound

Sierra Ladrones, north side

of treasure still hidden here." Herbert Ungnade reports *La Cueva de Ladrones*, "thieves' caves," southwest of Ladron Peak as a bandits' refuge.

Corry McDonald, in *Wilderness: a New Mexico Legacy*, retells one of the least apocryphal of the tales: "In 1910 a gang of robbers based in the Ladrones heisted a Wells Fargo shipment. A Wells Fargo bounty hunter spent two years trying to dislodge the robbers from the Ladrones before a final shootout on the Grants Malpais 50 miles away. Nothing was said about recovery of the bullion except that the case was closed."

Aside from treasure, human artifacts are few in the Sierra Ladrones, even place names. Despite scores of drainages, only two are named: Arroyo del Norte and Alamito Arroyo.

With perfect irony, the Ladrones, thought to contain so much treasure, are likely barren of precious metals. As one geologist summarized: "So far as this writer knows, only two metalliferous deposits have been found in the entire Sierra Ladrones district. At each of these copper is the chief metal, and neither constitutes an obviously workable deposit" (Samuel G. Lasky, *The Ore Deposits of Socorro County*). But he added that rugged terrain and ·
arduous access also have discouraged prospecting.

West summit, Ladron Peak

Geologically, there's no reason the Ladrones should not have mineralized veins, and in fact, a few veins of amethyst are found in the mountains. The small but compact range is basically a large uplifted mass of Precambrian granite and metamorphic rocks, with some Mississipian and Pennsylvanian sedimentary strata tilted on the western slopes. Similar configurations have yielded ore—though typically not much—elsewhere along the Rio Grande Rift. Geologically, the southeast foothills of the Ladrones are as complex as the rest of the range is simple. In the southeast are many small slides of rocks ranging in age from the very old Paleozoic through the much more recent mid-Cenozoic.

The few hikers who visit the Ladrones are those who accept—even welcome—the rugged terrain, and making the east-west traverse over the range is considered a cachet of hardiness. Perhaps the least difficult route to the top is from the west, where a dirt road leaves the graded county road and heads toward the mountains past a line camp. Hikers leave the dirt road, using map and compass or GPS to tell them where, and then go cross-country up a steep, rough gulch. According to Tom Ferguson, longtime NM Mountain Club member and hiker in the Ladrones, this was the route taken in 1952 by the newly formed NM Mountain Club on their inaugural hike as a club. Since then, Mountain Club members have hiked all over the world; google "New Mexico Mountain Club" to find the club's website.

In addition to the route above, some hikers follow the long northwest ridge to the top, a route much longer and rougher—and very exposed.

Even the insects in the Ladrones are prickly and unwelcoming. According to geologist Dave Love, "The Ladrones have an infamous cloud of voracious gnats that attack humans during May and June." Facetiously he adds, "They ate all the cattle and thieves" (personal communication).

Testifying further to the wild, rugged, and seldom-visited character of the Ladrones is the wildlife that lives here: mule deer, pronghorn, a reintroduced herd of bighorn sheep (1992), mountain lions, and black bears. The BLM Sierra Ladrones WSA encompasses 45,308 acres, but it also abuts the 228,000-acre Sevilleta National Wildlife Refuge, part of a nationwide, long-term ecological research program. The area here is closed to the public. (See *Los Pinos Mountains.*) Thus, a Sierra Ladrones Wilderness would be an important component of a much larger effective wilderness, one of great significance to central New Mexico.

Socorro, Chupadera, Lemitar Mountains

Making up a system of three small, closely related mountain groups, aligned north-south along the west side of the Rio Grande Valley, these mountains share a common—and complex—geological origin. From 75 to 40 million years ago, uplift created mountains here, followed by erosion down to a Precambrian core, with skiffs of Paleozoic rocks left behind. The time from 40 to 35 million years ago saw arcs of stratovolcanoes (imagine Mount Fuji or the volcanoes of the Cascade Mountains). The stratovolcanoes were succeeded 32–27 million years ago by supervolcanoes. These collapsed into giant calderas, the eroded rims of which coincided with the later Socorro and Chupadera Mountains. Since then, the mountains have been influenced by the development of the Rio Grande Rift. Rift-associated volcanism led to lava flows and volcanic domes along the Rift basin, creating Socorro Peak and Strawberry Peak. Finally, after 7 million years ago, continued uplift along the Rio Grande Rift, including the uplift of new fault blocks, led to today's configuration of the three ranges.

Socorro Mountains

Location: Immediately west of Socorro, north of US 60, south of
  Nogal Canyon.
Physiographic province: Basin and Range—Mexican Highland Section.
Elevation range: 4,900–7,000 feet.
High point: Unnamed summit 0.4 miles northwest of Socorro Peak,
  7,284 feet.
Other major peaks: Socorro Peak, 7,243 feet.
Dimensions: 3 x 6 miles.
Ecosystems: Chihuahuan Desert grasses and shrubs, piñon-juniper,
  mesquite, and creosote bush.
Counties: Socorro.
Administration: State (NM Institute of Mining and Technology), private.
Getting there: A dirt road leads from Socorro through Blue Canyon
  into the southeast part of the range. From Blue Canyon another
  dirt road branches north and leads to radio towers and research
  facilities atop Socorro Peak; access to this road is restricted by
  NM Institute of Mining and Technology.

Probably no New Mexico mountains have been as intensely studied as the

Socorro Mountains. Not only were they scoured exhaustively for minerals by prospectors in the late nineteenth century and earlier, but also they have been intensely examined by scientists with the NM Bureau of Geology and Mineral Resources, founded as an adjunct to the mining industry. Most of the mountains are now owned by NM Institute of Mining and Technology, which uses the mountains as an explosives research laboratory.

The Socorro Mountains are a small group, separated from the Chupadera Mountains on the south by Socorro Canyon and US 60, and from the Lemitar Mountains on the north by Nogal Canyon. Like their sibling mountains, the Socorro Mountains are comprised of a Precambrian igneous core overlaid by sedimentary formations, later capped by much more recent volcanic deposits. The faulting along the Rio Grande Rift that uplifted this block is best seen on the steep scarp on the east face of Socorro Peak.

It was this volcanism that created the mountains' silver and lead deposits. Though early Spanish workings have been reported, extensive mining did not begin until 1867, when English-speaking prospectors from Magdalena discovered commercially viable mineral deposits. Silver was the main ore, and the Merritt-Torrance vein, the district's richest, averaged fifteen to twenty ounces of silver per ton. Mining began in earnest in the 1880s, and soon three smelters were operating nearby. In 1889, the NM School of Mines, now NM Institute of Mining and Technology, was established. In a gesture that would be understood at colleges everywhere, mining students painted a huge M on the slopes of Socorro Peak, which is still referred to locally as M Mountain. Falling silver prices in the 1890s aborted most mining, though sporadic operations continued into the 1920s.

Blasting still occurs in the Socorro Mountains, however, for research rather than mining. Near Socorro Peak is a thirty-two-square-mile field laboratory owned by NM Tech and used by the Energetic Materials Research and Testing Center to study explosives. This has not prevented a herd of desert bighorn sheep from living in the mountains. The center annually makes accommodations for the Elfego Baca Shoot, a long-distance golf event in which golfers tee off from the summit of Socorro Peak and end at a designated hole on the NM Tech golf course. Average distance to a hole: 2.5 miles. The peak also is the focus of the annual M Mountain Duathlon, a forty-kilometer run and mountain-bike race around the mountain.

Geologically, the mountains are very young—and still growing. Beneath the town of Socorro is the Socorro Midcrustal Magma Body, which has made the area the most seismically active site in New Mexico during the past hundred years. Indeed, the strongest earthquake recorded in New Mexico, 6.5 on the Richter scale, occurred here in 1906. The magma chamber is about twelve miles beneath the Rio Grande Valley and its surface is rising about 0.08 inches per year. At that rate it will reach the surface in seven hundred twenty-six thousand years—making the Elfego Baca Shoot even more interesting.

Chupadera Mountains

Location: West of San Antonio, west of I-25, south of Socorro Canyon.
Physiographic province: Basin and Range—Mexican Highland Section.
Elevation range: 4,500–6,000 feet.
High point: Table Mountain, 6,300 feet.
Other major peaks: Chupadera Mountain, 6,273 feet.
Dimensions: 5 x 15 miles.
Ecosystems: Chihuahuan Desert grasses and shrubs, including mesquite and creosote bush.
Counties: Socorro.
Administration: Private, state, BLM—Socorro Field Office, U.S. Fish and Wildlife Service.
Wilderness: The mountains encompass the 5,289-acre Chupadera Wilderness, administered by Bosque del Apache National Wildlife Refuge and the U.S. Fish and Wildlife Service.
Getting there: Limited vehicular access. A dirt road that begins at the "Box" a few miles west of Socorro off of US 60 leads toward the mountains. See trail below.

The topographic complexity—hills, mesas, peaks—of the Chupadera Mountains is a reflection of their tortured geology. They, along with the Socorro Mountains immediately to the north across Socorro Canyon, are the eroded remains of volcanoes, which laid down thick layers of ash and lava atop Precambrian granite and later sedimentary layers. Collapse of the volcanoes into calderas also occurred here. Erosion has smudged the faulted and tilted volcanic shapes, but the Chupadera Mountains still rise abruptly from the neighboring plains and Rio Grande basin.

The mountains' desert ecology contrasts starkly with that of neighboring Bosque del Apache. From a trailhead located a mile north of the refuge's visitor center, the 4.8-mile Chupadera Mountain Trail leads into the Chupadera Wilderness, one of three wildernesses administered by the U.S. Fish and Wildlife Service, the others being the Indian Wells and Little San Pascual Wildernesses. Otherwise, all hiking is cross-country, through parched, complex terrain, though the major drainages of Walnut Canyon and Red Canyon slice through the Chupadera Mountains on their way east to the Rio Grande.

The Spanish name *Chupadera* means "sinkhole," but it's unknown whether the sinkhole was associated with the mountains first or with another feature, such as Chupadera Spring.

Lead and silver were mined at the range's north end, but operations ceased with the collapse of silver prices in the 1890s. More recently, manganese was produced from both open pit and underground mines during both World Wars and until the early 1970s in the Luis Lopez Mining District.

Lemitar Mountains

Location: About four miles west of the villages of Lemitar and Polvadera,
   north of the Socorro Mountains, south of the Sierra Ladrones.
Physiographic province: Basin and Range—Mexican Highland Section.
Elevation range: 5,200–6,000 feet.
High point: Polvadera Mountain, 7,292 feet.
Other major peaks: Strawberry Peak, 7,039 feet; Red Mountain, 6,698 feet.
Dimensions: 5 x 11 miles.
Ecosystems: Chihuahuan Desert grasses and shrubs, including mesquite
   and creosote bush.
Counties: Socorro.
Administration: BLM—Socorro Field Office, state, private.
Getting there: See below.

To persons driving on I-25 north of Socorro, the Lemitar Mountains appear as drab, brown bumps on the western skyline, dissected by numerous arroyos separating rounded, treeless ridges. Indeed, the Spanish name of the range's highest point, *Polvadera* Mountain, means "dusty," though that name originally was given to the village to the east, not the mountain. The range is bounded on the north by *Cañoncito de las Cabras* ("little canyon of the goats") and by San Lorenzo Arroyo. In the south the range ends at Nogal Arroyo. Unlike the Chupadera and Socorro Mountains to the south, prospectors found little of interest here beyond a few minor lead deposits.

Nonetheless, when Earth's crust cracked and heaved during the spreading of the Rio Grande Rift, it allowed numerous, small, carbonate dikes to pierce the underlying Precambrian granite and overlying Permian sedimentary formations. Especially on the range's western flank these dikes are associated with numerous, small, mineral deposits, of little interest to miners but of great interest to rockhounds. These deposits are primarily galena, sphalerite, barite, and bastnaesite, though wulfenite, hemimorphite, fluorite, cerussite, and others also are present.

Searching out these deposits is not easy, as no trails exist in the mountains, and steep slopes of decomposing granite can make hiking difficult. The route followed by the NM Mountain Club on their hikes up Polvadera Peak is to take the Lemitar Exit on I-25, and make a sharp right onto a frontage road heading north. Drive on this for about a mile to a metal gate on the left, just past a dip in the road. This dirt road (please close gate) goes 0.2 miles to a fork. Take the right branch for 3.3 miles to an informal parking area. From here it's a steep two-mile, cross-country hike to the summit.

Salado Mountains

Location: Approximately 11 miles west of Williamsburg, near Truth or
  Consequences, south of Salado Creek.
Physiographic province: Basin and Range—Mexican Highland Section.
Elevation range: 5,400–6,400 feet.
High point: Unnamed summit, 6,548 feet.
Other major peaks: None named.
Ecosystems: Piñon-juniper, alligator juniper, Chihuahuan Desert grasses
  and shrubs including mountain mahogany, yucca, and cacti.
Counties: Sierra.
Administration: BLM—Las Cruces Field Office.
Getting there: Consult the BLM office for current access; respect gates,
  signs, and private land.

A low group of hills, vegetated primarily by desert shrubs, the Salado Mountains are part of the Lake Valley string of uplifts that also includes Sierra Cuchillo, Sibley Mountain, and Animas Mountain, aligned north and south along the west side of the Rio Grande. Prospectors found fluorite here, but mining was minor and ceased long ago. The mountains' name, meaning "salty," likely evolved from the mountains' association with Salado Creek.

Sierra Cuchillo

Location: Northwest part of Sierra County, from Chise in the south to the Monticello Box in the north.

Physiographic province: Basin and Range—Mexican Highland Section.

Elevation range: 5,800–7,500 feet.

High point: Jaralosa Mountain, 8,326 feet.

Other major peaks: Reilly Peak, 8,167 feet; Cuchillo Mountain, 7,871 feet.

Dimensions: 9 x 16 miles.

Ecosystems: Piñon-juniper, Gambel oak, mountain mahogany, Upper and Lower Sonoran shrubs and grasses.

Counties: Sierra, Socorro.

Administration: Mostly BLM—Las Cruces Field Office.

Getting there: The best access is via NM 52, which bisects the range just before reaching Winston from Cuchillo.

Narrow and elongate, the Sierra Cuchillo is part of an almost continuous chain of hills running north from the Lake Valley Hills through the Animas Hills near Hillsboro and then northward, all expressions of a fault-block uplift along the Rio Grande Rift. In the Sierra Cuchillo, the escarpment faces west and the eastern slopes are much gentler. The range runs north, breached approximately in the middle by the gap at Red Hill—Schoolhouse Canyon, through which NM 52 runs.

Prospectors discovered mineralization—iron, copper, and lead-zinc— in contact-metamorphism deposits at the north end of the range. At about the same time much richer deposits were discovered at nearby Winston and Chloride. Conflict with resident Apaches delayed mineral development until 1900, but even then mining never was significant. It did, however, have the effect of removing most of the ponderosa pines from the range's north-ern upper slopes, which were cut for mining development. Today piñon-junipers and Chihuahuan Desert shrubs dominate the vegetation.

It would be tempting to conclude that the Spanish name *Sierra Cuchillo,* "knife range," was a descriptive metaphor, but as *The Place Names of New Mexico* explains, it refers instead to *Cuchillo Negro,* "Black Knife," the leader of the Warm Springs Apache band whose territory was here. Numerous features in his former homeland bear his name.

Mud Springs Mountains

Location: Immediately northwest of Truth or Consequences.
Physiographic province: Basin and Range—Mexican Highland Section.
Elevation range: 4,500–5,500 feet.
High point: Mud Mountain, 5,749 feet.
Other major peaks: None.
Dimensions: 2.5 x 5 miles.
Ecosystems: Scattered piñon-juniper, but mostly Chihuahuan Desert
    grasses and shrubs, including creosote bush, ocotillo, yucca, and cacti.
Counties: Sierra.
Administration: BLM—Las Cruces Field Office.
Getting there: A few dirt roads and ATV tracks head into the mountains
    from Truth or Consequences.

The Mud Springs Mountains are a linear group of small, slender southeast-northwest-trending ridges, carved by erosion out of a fault block tilted toward the northeast. The rounded slopes—gentler east, steeper west—and the sinuous pattern of eroded, Paleozoic, sedimentary layers evoke images of a giant serpent burrowing in the exposed landscape. Prospectors once found minor copper and silver deposits here, but today the desert range is largely ignored, and cacti, agaves, and ocotillo show more color than anything in the ground. (Note: This is not to be confused with Mud Spring Mountain and Mud Spring farther west, in the eastern foothills of the Black Range.)

Tonuco/San Diego Mountain

Location: East of the Rio Grande, just west of I-25, 25 miles north of
    Las Cruces.
Physiographic province: Basin and Range—Mexican Highland Section.
Elevation: 4,949 feet.
Relief: 950 feet, above the Rio Grande floodplain to the west.
Ecosystems: Lower Sonoran grasses and shrubs, including grama grasses,
    creosote bush, mesquite, yucca, and cacti.

Counties: Doña Ana.

Administration: BLM—Las Cruces Field Office.

Getting there: A jeep road leads to the summit after branching west from a
    dirt road paralleling I-25 on lands of the NM State University Animal
    Science Ranch.

Spaniards traveling up the Rio Grande along the Camino Real passed
beneath the hulking mass of Tonuco Mountain. They called it "San Diego,"
for San Diego *paraje,* or campsite, just beneath the mountain at an important river ford. Yet the name Tonuco, of unknown origin and meaning,
appears on Spanish maps as early as 1828 and now seems to be supplant-
ing San Diego on most new maps. The smaller formations around the main
mass are called the Tonuco Mountains.

    The mountain has a Precambrian core, overlain by Paleozoic and
Cenozoic sandstones, but it didn't attain its present form until much later,
during the late Cenozoic, when eight thousand feet of volcanic rocks were
deposited and the whole mass faulted and arched upward to become the
highest point on the three-mile-long Tonuco Uplift. Sometime during all of
this, barite and fluorite were deposited, and between 1919 and 1921 about
two thousand five hundred tons of ore were shipped from mines on the
mountain's southeast side. The minerals proved difficult to extract, however, and mining ceased long ago. Now the main activity on the mountain
is jeeping over abandoned mining and jeep roads.

Sierra de las Uvas

Location: Approximately 28 miles northwest of Las Cruces, west of
    Radium Springs, south of Hatch.

Physiographic province: Basin and Range—Mexican Highland Section.

Elevation range: 5,000–6,200 feet.

High point: Magdalena Peak, 6,625 feet.

Other major peaks: Sugarloaf Peak, 6,494 feet; Tailholt Mountain,
    6,027 feet; Pina Peak, 5,949 feet.

Dimensions: 10 x 20 miles.

Ecosystems: Chihuahuan Desert shrubs and grasses including grama,
    mesquite, creosote bush, sotol, scrub oak, and juniper.

Counties: Doña Ana.

Petroglyphs, Sierra las Uvas

Administration: BLM—Las Cruces Field Office.
Wilderness: 11,067 acres in the range's northern section make up the
    BLM Las Uvas Mountains Wilderness Study Area.
Getting there: From NM 185 about halfway between Radium Springs
    and Hatch take dirt County Road E006 west. After about 12 miles
    is the WSA's southern boundary, evidenced by high mesas to
    the north.

The *Sierra de las Uvas* is a sprawling mass of high mesas and peaks, north
of the Robledo Mountains but geologically distinct from them. The name
means "range of the grapes," but as *New Mexico's Wilderness Areas* and *The
Place Names of New Mexico* explain, don't look for *uvas* in the Sierra de las
Uvas. Two explanations exist for the name. One says it referred to a huge,
wild grape thicket at a spring in the mountains, where a single plant once
covered more than an acre. But a geologist has said the uvas were
spheroidally-weathered, purplish-gray ovoids of basaltic andesite, reminiscent of giant grapes (personal communication). Rather, look for barrel

cacti, ocotillo, prickly pear, mesquite, creosote bush, and other plants of the Chihuahuan Desert. Raptors, including golden eagles, nest in vertical, basaltic cliffs capping the higher mesas.

The mountains were formed in a process that began 35–28 million years ago. Then, tuff from volcanic vents was deposited in an adjacent subsidence structure known as a *half graben*. By 28–27 million years ago this had become an early basin of the Rio Grande Rift. As the rift continued to develop with subsidence shifting to the east, the formations to the west, originally created in a basin, were uplifted into an arch in a process known as structural reversal. About 10–15 million years ago, the former half-graben layers were broken, tilted, uplifted, and eroded, leading to today's topography of the Sierra de las Uvas.

Decades of overgrazing have contributed to the mountains' present scraggly vegetation and barren demeanor, but the mountains were not always thus. Petroglyphs throughout the mountains, but especially in Broad and Valles Canyons, speak of Indians of the Jornada Mogollon Culture (AD 850–1350) living here. In Broad Canyon they pecked the shape of a fish— intriguing, as today drainages in the Sierra de las Uvas usually don't even have water, much less fish.

The Sierra de las Uvas have no developed hiking trails and attract few hikers, though cross-country hiking here is straightforward over the open terrain, laced with numerous ranch roads and tracks also used by off-road vehicles. Despite these incursions, the mountains retain their wild character and, along with the neighboring Robledo Mountains, are considered a candidate for formal wilderness designation and protection.

Robledo Mountains

Location: Immediately west of the Rio Grande, 8 miles northwest of Las Cruces and south of Radium Springs, from Faulkner Canyon to Picacho Mountain.

Physiographic province: Basin and Range—Mexican Highland Section.

Elevation range: 3,950–5,500 feet.

High point: Robledo Mountain, 5,890 feet.

Other major peaks: Lookout Peak, 5,648 feet; Picacho Mountain, 4,959 feet.

Dimensions: 5 x 11 miles.

Ecosystems: Chihuahuan Desert grasses and shrubs, including grama grasses, tobosa grasses, creosote bush, mesquite, mountain mahogany, yucca, and cacti, with scattered junipers and oaks at higher elevations.

Counties: Doña Ana.

Administration: BLM—Las Cruces Field Office.

Wilderness: The BLM Robledo Mountains Wilderness Study Area, 12,496 acres, is mostly on the range's western side.

Getting there: Approach the Robledo Mountains from the northwest by taking NM 185 north through Radium Springs, continuing 0.5 mile west of the Blue Moon Bar and Cafe then turning south onto dirt Faulkner Canyon Road (County Road D59). Approach the mountains from the southwest by taking the I-10 frontage road marked "Airport" at the US 70 and I-10 junction west of Las Cruces, then taking Corralitos Road (County Road C09) north about 7 miles to where County Road C07 branches northeast, leading to the mountains.

On May 21, 1598, as New Mexico's first European colonists moved up along the Rio Grande led by Don Juan de Oñate, the expedition mourned its first death when sixty-year-old Pedro Robledo died of natural causes. He left behind his name on the mountains overlooking his unmarked grave.

And that alone has for hundreds of years kept the Robledo Mountains from complete obscurity, eclipsed by the far more dramatic Organ Mountains just to the south. The Robledos are a low range of unprepossessing contours, with relief less than two thousand feet. For most of the year the mountains are exposed, searing hot, and dry. Only a few old dirt tracks—and some recent illegal ones—lead into the range. There are no developed hiking trails, though cross-country travel is straightforward. Most people view the Robledos from I-25 and see them as a rather attractive tableau of sinuous sedimentary layers, mostly Paleozoic limestone. And it was those layers that finally gave the mountains a significance beyond the old Spaniard.

Paleontologists long had recognized the Robledos, a southwest-tilted fault block of mostly Pennsylvanian to Permian limestone and shales, as an important research area for both marine and terrestrial sedimentology, but in 1987 paleontologist Jerry MacDonald, searching a terrestrial layer between Permian limestone layers, found the footprints of amphibians and reptiles and even insects that had walked across the mud flat 280 million

Barrel cacti, Robledo Mountains

years ago. The site, now managed by the BLM, proved to be the world's finest fossil track site of its type. Elsewhere in the Robledos the rocks preserve the fossils of both marine and terrestrial invertebrate animals, as well as fossil plants.

Similarly, archaeological sites in the Robledos, while small and inconspicuous, nonetheless are significant. A Folsom point was found in a cave site, rare evidence of Paleo-Indian presence in southern New Mexico.

The Robledos have other attractions as well. Because the area does indeed have wilderness characteristics, it's a good place to explore desert ecology. Barrel cacti are common here, as well as other cactus species. Hikers bound for the highest point, Robledo Mountain, follow an unmarked path that begins on Faulkner Canyon Road, goes south up an arroyo, then links with a 4WD road heading east and up between Lookout Peak and Robledo Mountain. At a road fork, the track to the right leads to Robledo Mountain.

Picacho Mountain (Doña Ana)

Location: Northwest of the city of Las Cruces, toward the south end of
the Robledo Mountains.
Physiographic province: Basin and Range—Mexican Highland Section.
Elevation: 4,959 feet.
Relief: 900 feet.
Ecosystems: Chihuahuan Desert shrubs and grasses.
Counties: Doña Ana.
Administration: Private.
Getting there: The peak is reached from Picacho Ave. heading west
from Las Cruces.

Viewed from several directions this conspicuous cone is a local landmark. The name Picacho Mountain is one of those oxymorons common in bilingual areas. It consists of two generic terms, one Spanish and the other English, both meaning approximately the same thing.

Structurally, Picacho Mountain is connected by low ridges with the Robledo Mountains seven miles north. Geologically it is a flow banded rhyolite dome, about 35 million years old. Picacho's conspicuity has spawned local legends of treasure, but except for some hematite deposits in the former Iron Hill Mining District, no significant mineralization has been found. The Butterfield Overland Mail route passed just to the north of Picacho Mountain, in Apache Canyon.

East Potrillo Mountains

Location: Just north of the Mexican border, southwest of Las Cruces.
Physiographic province: Basin and Range—Mexican Highland Section.
Elevation range: 4,200–5,100 feet.
High point: Unnamed summit, 5,359 feet.
Other major peaks: None.
Dimensions: 1 x 7 miles.
Ecosystems: Mesquite, creosote bush, manzanita, Chihuahuan desert
    grasses, and widely scattered junipers and scrub oak.
Counties: Doña Ana.
Administration: BLM—Las Cruces Field Office, state.
Getting there: The BLM has published a menu of county roads leading
    to Kilbourne Hole; from there, a county road continues west, then
    southwest to pass between the East Potrillo Mountains and Cox
    Peak/Mount Riley. The BLM 1:100K Surface Management Status
    map is the best depiction of the myriad dirt roads in this area.

The East Potrillo Mountains, unlike the longer but more diffuse West Potrillo chain of volcanoes, are a compact, narrow ridge of mostly sedimentary rocks running northwest-southeast beginning just three miles north of the Mexican border. They rise almost 1,000 feet above the surrounding basin, or bolson, but they're overshadowed by much-higher Cox Peak (5,957 feet) and Mount Riley (5,905 feet) (see *West Potrillo Mountains*) immediately northwest. No trails lead into the mountains, and except for occasional hunters and rockhounds, few people come here.

Geologists are fond of the East Potrillos, however, because the steep slopes provide superb exposures of the sedimentary rocks, mostly limestone and sandstone that make up the mountains. These are folded and thrust eastward over each other in a complex jumble, then tilted southwest. As the geologists William R. Seager and Greg H. Mack describe the mountains, "Although the relief here is low, the land is nevertheless rugged in its own way, owing to the youthfulness of the flows and to the uninhabited, desolate tracts of sand-strewn malpais" (*Geology of East Potrillo Mountains and vicinity, Doña Ana County, NM*). (For name and ecological information, see *West Potrillo Mountains.*)

Actually, the most interesting features in this area are not above the ground but in it. They are maars, shallow, flat-floored volcanic craters formed from explosions that ejected mostly gas. Three large maars—Kilbourne Hole, Hunts Hole, and Potrillo Maar—exist north, east, and south of the East Potrillo Mountains, but Kilbourne Hole—1.7 miles long, 1 mile wide, and several hundred feet deep—is the most impressive. Chunks of Earth's mantle caught in the rising magma and ejected in the volcanic shrapnel commonly contain peridot, the bright green gemstone.

West Potrillo Mountains

Location: Southwest of Las Cruces, extending from south of I-10 to just north of the Mexican border.

Physiographic province: Basin and Range—Mexican Highland Section.

Elevation range: 4,500–5,100 feet.

High point: Potrillo Peak, 5,397 feet; Cox Peak, 5,957 feet; Mount Aden, 4,709 feet.

Other major peaks: Mount Riley, 5,905 feet; Guzmans Lookout Mountain, 4,762 feet.

Dimensions: 7 x 28 miles.

Ecosystems: Mesquite, creosote bush, manzanita, Chihuahuan desert grasses, and widely scattered junipers and live oak.

Counties: Doña Ana.

Administration: BLM—Las Cruces Field Office, state.

Wilderness: The BLM Aden Lava Flow, West Potrillo Mountains, and Mount Riley Wilderness Study Areas include 159,972 acres.

Getting there: As with the East Potrillo Mountains, the BLM has published a menu of county roads leading to Kilbourne Hole; from there, a county road continues west, then southwest to pass between the East Potrillo Mountains and Cox Peak—Mount Riley. The BLM 1:100K Surface Management Status map is the best depiction of the myriad dirt roads in this area. The Aden features and also the West Potrillo Mountains can be approached by taking County Road 4 south from NM 549, east of Exit 116 on I-10.

Geologists William R. Seager, John W. Hawley, and Russell E. Clemons summarized the West Potrillo Mountains especially well:

> The West Potrillo Mountains are an uninhabited stretch of
> uplands lying just west of the East Potrillo Mountains. . . .
> Rather than consisting of deformed and uplifted rock, the
> West Potrillo 'Mountains' are a collection of more than 150
> cinder cones, maar volcanoes, and voluminous basalt
> outpourings, extensively covered by a thin mantle of sand
> but still rugged enough and desolate enough to compare
> favorably with a lunar landscape. (*Geology of San Diego
> Mountain Area, Doña Ana County, NM*)

Certainly, looking down and west from atop Cox Peak evokes cratered desolation. The West Potrillo Mountains are a chain of volcanic cysts, ranging from 3 million years old to 0.5 million years ago, in the main part of the volcanic field. Little Black Mountain is around two hundred sixty-eight thousand years old, while Aden Crater is probably as little as twenty thousand years old. These dark-hued, volcanic features squat upon an otherwise featureless landscape of dark-green creosote bush, unrelieved by any brighter green betokening moisture.

It's every bit the wild land that wilderness advocates claim it to be. Stretching approximately twenty miles from the Aden Hills and Aden Cone on the north to Guzmans Lookout Mountain in the south, the West Potrillo Mountains are possibly New Mexico's longest inconspicuous mountain range. The WSA here is the state's largest. Animals have made the trails here, though humans have created the many dirt roads accessing the area. One hiker had a very satisfying backpack going from north to south, during one of the rare periods when snow hung like lace over the landscape. He found seven cones whose craters were unbreached.

The dominating summits of Cox Peak and Mount Riley are the most obvious destinations for visitors. They're included here with the West Potrillo Mountains not by proximity—they actually are closer neighbors of the East Potrillos—but because like the West Potrillos they were born as igneous extrusions, rhyodacite in the case of Cox and Riley. They are the remains of a pluton (an igneous intrusion) that became emplaced here in the early to mid-Cenozoic, 32–26 million years ago. Today they rise as stark anomalies

East flanks of Cox Peak

above the much lower, much more recent volcanic formations of the West Potrillo Volcanic Field. A hiker attempting to climb them confronts very steep, sparsely vegetated scree slopes of flat, conchoidally-fractured fragments that clank and slip like broken plates. The east ridges offer the least difficult approaches, and the hard-won summits offer compensatory views.

This volcanism also created the features surrounding the railroad siding of Aden: the Aden Hills, 2.5 miles north of the railroad, highest elevation 4,780 feet; Aden Cone, also called Mount Aden, 4,709 feet; Aden Crater, 4,477 feet, 6.75 miles southeast of Aden; and the Aden Lava Flow. Except for their names, they would not be distinguished from dozens of other volcanic cones and craters along a fifty-mile axis running from the Rough and Ready Hills south to the Mexican border. Aden Crater attracted attention when a mummified ground sloth was found in a shallow cave in the crater.

Soaptree yuccas and Mount Riley

Hunters come here for deer and quail. And the austere environment notwithstanding, wildlife here is relatively abundant: deer, badgers, bobcats, javelinas, quails, raptors, and many more. In 1900, forty desert bighorn sheep were killed here—eleven years after bighorn hunting was prohibited. None have been reported since.

The vegetation also exemplifies the Chihuahuan Desert: creosote bush, mesquite, sotol, snakeweed, soaptree yucca, and cacti including prickly pear and barrel. A few junipers are scattered among the mountains. Missing from the fauna, however, are the wild horses, whose offspring, *potrillos,* "colts," likely inspired the mountains' name. And there are only a few ranches. As Mike Hill said in his *Guide to the Hiking Areas of New Mexico,* "There are probably acres in the West Potrillo Mountains that have seen very few footprints since prehistoric times. If solitude is what you want, this is an extreme example of it."

Good Sight Mountains—Nutt Mountain

Location: Northeast of Deming, north of I-10, near NM 26.
Physiographic province: Basin and Range—Mexican Highland Section.
Elevation range: 4,140–5,600 feet.
High point: Nutt Mountain, 5,940 feet.
Other major peaks: Good Sight Peak, 5,602 feet.
Ecosystems: Creosote bush, mesquite, saltbush, and desert grasses.
Counties: Doña Ana, Luna, and Sierra.
Administration: Mostly BLM, some state and private.
Getting there: Access is via county roads branching south from NM 26.

The Good Sight Mountains, like the related Sierra de las Uvas to the east across the Uvas Valley, are primarily volcanic remnants. About 30 million years ago, during the widespread volcanism of the Oligocene Epoch, a volcanic vent zone ran north from I-10 about forty miles to Nutt Mountain. When the eruptions stopped, warping, faulting, and erosion began sculpting the landscape into the low, sparsely vegetated hills we see today. From the east the slopes are gentle; the western escarpment is somewhat bolder. To the north are cuestas (linear ridges) of ashfall tuff. The mountains' volcanic birth resulted in numerous dikes, plugs, stocks, lava flows, and other intrusive structures comprised of such volcanic rocks as andesite, latite, rhyolite, and tuff. Nutt Mountain itself is an intrusive, flow-banded, rhyolite plug.

Hiking here is restricted to cross-country travel and ranch roads. Rather, the Good Sight Mountains are primarily grazing land. The grass here certainly would have been valued by the Butterfield Overland Mail, for they established a relay station here, at Good Sight Spring. From this the mountains' name likely was created.

The range's most conspicuous—and highest—summit is Nutt Mountain. The jutting core of a long-extinct volcano, rising abruptly from the nearby plains, Nutt Mountain is an obvious landmark for the widely-scattered inhabitants and travelers in these parts. It certainly would have been familiar to the miners, prospectors, freighters, and other boomtown entrepreneurs who flocked to the area when silver was discovered at Lake Valley in 1878. The rail line between Nutt and Lake Valley ran just beneath Nutt Mountain. Colonel Nutt was a railroad director. The peak also has been called Sunday Cone, perhaps a misspelling of Sundae Cone, a plausible metaphor for the formation.

Pinos Altos Range

Location: North of Silver City, south of Sapello Creek and the Gila River, west of the Mimbres River.

Physiographic province: Basin and Range—Mexican Highland Section.

Elevation range: 5,200–8,200 feet.

High point: Black Peak, 9,209 feet.

Other major peaks: Signal Peak, 8,980 feet; Tadpole Ridge, 8,614 feet; Twin Sisters, 8,347 feet north, 8,196 feet south; Pinos Altos Mountain, 8,116 feet.

Dimensions: 10 x 20 miles.

Ecosystems: Piñon-juniper, mountain mahogany, ponderosa pine, Douglas fir, western white pine.

Counties: Grant.

Administration: Gila National Forest—Silver City Ranger District.

Getting there: Paved access to the range's eastern reaches is via NM 15 from Pinos Altos and NM 35 from Mimbres. Access to the west is from Silver City via paved Little Walnut Road/dirt Forest Road 506 or US 180.

The Pinos Altos Range is a poorly defined group of mountains and hills arching between the Gila River, Sapello Creek, the Mimbres River, Silva-San Vicente Creeks, and Whitewater Creek-Lampbright Draw, hardly distinguished topographically from its neighbors to the southwest, the Silver City Range and, still farther, the Burro Mountains. All share a geologic history based upon the general mid-Cenozoic volcanism of the Datil-Mogollon Section of the Transition Zone. Most of the range here is composed of ashfall tuffs that accumulated between 35.2 and 28 million years ago. Much of the ashfall came from the volcano whose 31.4 million-year-old caldera is centered on the Twin Sisters peaks.

In the Pinos Altos Range this geology became much more interesting, at least to modern humans, when mineralization occurred depositing gold and silver. In 1856 a rich placer gold deposit was discovered on Bear Creek near Pinos Altos; it was so rich that miners could make forty to fifty dollars a day panning, a staggering amount at the time. Within three months a thousand prospectors had flocked to the area, and the Bear Creek Mine became the first corporate mining company in New Mexico. But persistent Apache attacks, including major raids in 1861 and 1864, aborted the boom and led to the temporary abandonment of the camp, then called Birchville, for Thomas "Three Finger" Birch, one of the three prospectors who had made the original strike. It is said that fleeing miners looked back to see Apaches torching their former homes. When the abandoned camp was resettled, it took the name by which it earlier had been known to Spanish-speakers, *Pinos Altos,* meaning "tall pines," for its ponderosas. Numerous arrastres, crude ore-crushing devices used by early Hispanic miners, have been found at several places in the Pinos Altos Range, and the mountains are littered with abandoned mines, prospect holes, mill sites, and mining roads.

Probably the best hiking route here is the Continental Divide Trail, which enters the range from Forest Road 506, about five miles north of Silver City. The pleasant and scenic route keeps close to the physical Divide all the way to the Mimbres River. Trails beginning on the Fort Bayard reservation also lead to the Divide. The mountains are uniformly dry, save for a few small and predominantly intermittent springs and streams. Thus the mountains support fairly uniform vegetation—ponderosa pines, piñon and alligator juniper, and live oak.

Silver City Range

Location: North and east of Silver City, from west of Chloride Flats,
    north of US 180, to west of NM 15.
Physiographic province: Basin and Range—Mexican Highland Section.
Elevation range: 6,200–7,900 feet.
High point: Bear Mountain, 8,036 feet.
Other major peaks: McComas Peak, 7,681 feet; Little Bear Mountain,
    7,505 feet.
Dimensions: 6 x 15 miles.
Ecosystems: Piñon-juniper, ponderosa pine, oaks.
Counties: Grant.
Administration: Gila National Forest—Silver City Ranger District.
Getting there: From Silver City, Forest Road 853—the Bear Mountain
    Road—goes north to Bear Mountain, while county-maintained
    Little Walnut Creek road goes to recreation sites farther east.

Dwarfed by its neighbors, the Big Burro Mountains and the Pinos Altos Range, the Silver City Range is redeemed from obscurity by Bear Mountain, located at the range's north end and a commanding landmark in the Silver City area. Only slightly less impressive is McComas Peak, immediately south of Forest Road 853. Both peaks have unmaintained trails leading to their summits from the Bear Mountain Road. The geology that created these mountains echoes that of ranges throughout the region: a northeast-tilted fault block of thick Paleozoic and Cretaceous sedimentary strata intruded by mid-Cenozoic igneous formations, here quartz monzonite porphyry stock near Silver City.

Whatever inspired the name Bear Mountain has been forgotten, but McComas Peak recalls a tragic incident from the Apache Wars still remembered in Silver City. In March, 1883, Judge H. C. McComas, his wife, Juniata, and their seven-year-old son, Charlie, were heading southwest from Silver City toward Lordsburg when their wagon was ambushed by Apaches. McComas and his wife were killed, which devastated Silver City where the judge was respected and well liked. But much worse was that Charlie was missing, likely abducted by the Apaches. For years rumors of the boy sent searches throughout the region, but he was never found. Years later, Apache sources confirmed that the boy had been a captive, and some say he died when an Apache, notorious for his bad temper, in a rage dashed the boy's brains out on a rock.

The Silver City Range, because of its proximity to Silver City, is popular with hikers, mountain-bikers, trail-runners, and equestrians. Also, the Silver City Range is aligned along the Continental Divide, and the Continental Divide Trail goes through here. The terrain is rolling and pleasant, characterized by ponderosa and piñon-juniper forests interspersed with meadows. It's good habitat for elk, deer, bear, wild turkey, and other large vertebrates. But the mountains are conspicuously dry. Little or no surface water exists along the CDT here.

Burro Mountains (Big and Little)

Location: About 20 miles southwest of Silver City, just north and
  south of NM 90.
Physiographic province: Basin and Range—Mexican Highland
  Section.
Elevation range: 4,200–8,000 feet.
High point: Big Burro Mountains, Burro Peak, 8,035 feet; Little Burro

Mountains, 6,637 feet, summit unnamed on most maps but said to
be called *La Lacha*, meaning obscure.

Other major peaks: Jacks Peak, 7,986 feet; Ferguson Mountain, 7,960 feet;
Bullard Peak, 7,064 feet; Hornbrook Mountain, 7,048 feet, highest
point in the southern Big Burro Mountains.

Ecosystems: Ponderosa pine, Chihuahuan pine, Gambel oak, and
mountain mahogany, as well as Emory, gray, and Arizona white
oak, galleta, and Indian ricegrass.

Counties: Grant.

Administration: Gila National Forest—Silver City Ranger District,
numerous private inholdings in the Big Burro Mountains; the
Little Burros are mostly private.

Getting there: The best access is from NM 90 southwest of Silver City and
US 180 northwest of Silver City; the good dirt Mangas Valley Road
connects the two as it goes past the Phelps Dodge mining operations.
From the paved roads several Forest Roads go into the mountains.
From NM 90, Forest Road 841 leads to the Gold Hill area and other
Forest Roads. From the Mangas Valley Road, Forest Road 851 bisects
the range on the north and connects with other Forest Roads.

Proximity and geology have linked these two ranges, both having an
igneous, Precambrian core and separated only by the Mangas graben, or
"Mangas trough," atop which lies the valley of Mangas Creek. The Little
Burro Mountains are indeed "little," only eight miles long, a northeast-tilted
fault block with relief of only one thousand two hundred feet on gently contoured tops. Motorists driving on NM 90 through the Little Burros rarely
have a sense of being in mountains at all. This and lack of public access
means the Little Burros will continue to be a mere footnote to the Big
Burros, except, of course, for the copper mines at Tyrone.

The Big Burros, twenty miles long and with relief of two thousand feet,
undisputedly are mountains. NM 90 divides the range into southern and
northern parts, the northern higher. The northern mountains are formed of
faulted, Precambrian granite and metamorphics, interspersed with Cretaceous and mid-Cenozoic volcanics and intrusions and Gila conglomerate.
The conglomerate rock resembles the compressed volcanic ash formations
so common in the region, but in fact the detritus here fell not from the sky
but was deposited when ancient streams around the Gila River's upper

drainage basin flooded (as streams here do today). Much of the conglomerate also resulted from an arid period when uplift had slowed and deep layers
of rocky sediments accumulated on pediments fringing the mountains'
foothills. South of the Tyrone Mine the geology is dominated by a late-
Cretaceous grandiorite intrusion. Burro Peak and Jacks Peak are Precambrian
granite. Throughout the range's central and southern mountains are mid-
Cenozoic granite and diorite dikes trending northeast-southwest.

Despite sharing the name Big Burro Mountains, the section of the Big
Burro Mountains generally south of NM 90 is geologically distinct from the
mountains to the north. The southern mountains consist of a tilted fault
block of Precambrian granite, later intruded by mid-Cenozoic dikes, and
finally overlain by tilted volcanics.

With this igneous activity came mineralization and, much later, mining. Today, with the Burros seldom visited except by a few hunters, four-
wheelers, dirt bikers, mountain bikers, and Continental Divide Trail hikers,
it's hard to imagine that slightly more than a hundred years ago the mountains were the focus of intense human interest as prospectors and miners
swarmed over the hillsides. As many as nine mining camps, many with post
offices, flourished here—briefly.

Long before Europeans arrived, prehistoric Indians, members of the
Mimbreño branch of the Mogollon Culture (AD 850–1350) mined turquoise
in these mountains. In 1875 John Coleman (later dubbed "Turquoise John")
rediscovered their ancient diggings west of Saint Louis Canyon. Other
turquoise mines were opened nearby; in the Azure Mine, a 40 x 60-foot vein
known as the Elizabeth Pocket produced $2 million in turquoise.

But the real mining boom of the 1870s came with copper, gold, and silver strikes. Most occurred toward the range's northern end, where mining
camps such as Telegraph, Paschal, Black Hawk, Oak Grove, Carbonate City,
Fleming, and Penrose sprang up. In the south were the gold camps of
Rattlesnake City and Gold Hill, while in the Little Burro Mountains was
Leopold, a copper camp. Bullard Peak, 7,064 feet, at the range's north end,
recalls U.S. Army Captain John Bullard, miner and prospector who around
1870–71 developed several claims around the peak now bearing his name.
In 1871, while leading a campaign against the Apaches, Captain Bullard was
shot and killed on another mountain, now also called Bullard Peak, west of
Glenwood, just over the Arizona line. By mid-1882 the town of Paschal had
approximately one thousand residents and was one of the nation's leading

copper camps. But after 1883, when copper prices fell and labor costs rose, the town began to die, and few of the camps survived into the twentieth century, though Leopold had one thousand two hundred residents in 1907, and Tyrone still is an active company town for the Phelps-Dodge Corporation's open-pit copper mines nearby. Today, most prospecting is by rockhounds seeking mineral specimens, such as crystal clusters of dark-purple fluorite found around Spar Hill.

There are no burros in the Burro Mountains. And according to *The Place Names of New Mexico,* "Whatever burros inspired this range's name must have lived a long time ago, because Juan Nentwig's map of 1762 shows *S. [Sierra] de las Burros* just west of *S. del Cobre* and south of the *Rio Xila.*" The animals most likely to be found here are white-tailed and mule deer, javelinas, black bears, coyotes, and mountain lions. A desert bighorn ram was killed here in 1900; one hundred years later, in 2000, sixty-five bighorn sheep were reintroduced to the mountains. In 1999 an unconfirmed but very credible sighting of a black-phase jaguar was reported. The habitat consists primarily of desert plants such as yucca, desert willow, mesquite, and creosote bush at lower elevations, and piñon-juniper and then ponderosa pine with increasing elevation. These species represent a significant recovery from severe overgrazing in the nineteenth and early twentieth centuries. A 1902 livestock survey showed two thousand sheep, one thousand five hundred cattle, and at least one hundred horses in the mountains that year.

Outdoor recreation in the Burro Mountains always will be overshadowed by the much better known opportunities in the nearby Gila Country. Moreover, the Burros, while pleasant, lack a specific natural or cultural focus of interest and for much of the year are dauntingly hot and dry; natural water sources are small and few. Still, hikers wanting to ascend the highest summits—and also to experience a section of the Continental Divide and the Continental Divide Trail—will find a pleasant day hike in Trail 74. The trailhead, well-marked and with ample parking—but no water—is about twenty miles southwest of Silver City on NM 90. The trail ascends steadily but not too steeply to Jacks Peak, 7,986 feet in elevation, topped by a thicket of transmission towers and offering spectacular vistas of southwestern New Mexico. Burro Peak, at 8,035 feet, is 0.75 miles farther north. Natural water sources are lacking throughout. Perhaps the best way to explore the Burros is by mountain bike or horse, and several routes are overlaid upon the network of old mining and logging roads.

Langford Mountains

Location: At the south end of the Big Burro Mountains.

Elevation range: 5,400–6,200 feet.

High point: Unnamed summit, 6,272 feet.

Ecosystems: Creosote bush, saltbush, Gambel oak, mountain mahogany, and desert grasses.

Counties: Grant.

Administration: Mostly private but also with BLM, state, and Forest Service land.

Getting there: Access is primarily via Forest Roads through the southern Burro Mountains, specifically Forest Road 841 south from NM 90.

The Langford Mountains dangle like an appendix from the much larger, more complex Big Burro Mountains. This linear desert range, a squiggle barely three miles long, is bounded on the north by Jones Canyon, on the south by Langford Draw. Despite limited access and an extremely steep and rough crest ridge, the Langford Mountains have been scouted as an alternative route for the Continental Divide Trail, and hikers who have explored here return enthusiastic about the scenery.

The mountains are a southeastern extension of the Precambrian granite that underlies the Burro Mountains to the north. The region has been marked by numerous intrusive formations, and to the east-northeast these have resulted in several conspicuous hills, including Soldiers Farewell Hill, Bessie Rhoads Mountain, and JPB Mountain.

Grandmother Mountain and Little Grandmother Mountain

Location: West of Deming, 6.4 miles north of Exit 62 at Gage on I-10 west of Deming.

Physiographic province: Basin and Range—Mexican Highland Section.

Elevation: Grandmother Mountain, 5,866 feet; Little Grandmother Mountain, 5,165 feet.

Relief: Grandmother Mountain, 500 feet; Little Grandmother Mountain, 400 feet.

Ecosystems: Chihuahuan Desert grasses and shrubs, including grama, mesquite, tarbush, and creosote bush.

Counties: Luna.

Administration: State, some private land around Little Grandmother
    Mountain.

Getting there: From the Gage Exit on I-10, a dirt road running north
    soon bifurcates into roads going both east and west around
    Grandmother Mountain.

Grandmother Mountain is an agglomeration of approximately fifteen latite
plugs and volcanic flows, overlying Cretaceous and early-Cenozoic sediments. The cones are a localized expression of the general volcanic upheavals in southwestern New Mexico about 30 million years ago. As for the
curious name, Grandmother Mountain, that remains a mystery.

Little Grandmother Mountain, one mile northeast of Grandmother
Mountain across Cow Springs Draw, is a cuesta, a narrow, linear ridge of
29-million-year-old ashflow tuff, 3 miles long and shaped like a shepherd's
crook when seen from above or on a map. The formation rises gradually
from the east to an abrupt and very steep western escarpment.

A related feature is Clabber Top Hill, five thousand one hundred feet in
elevation, a system of at least two latite plugs, about five miles northeast of
Grandmother Mountain.

Cookes Range

Location: North of Deming, north of NM 26.

Physiographic province: Basin and Range—Mexican Highland Section.

Elevation range: 5,250–7,600 feet.

High point: Cookes Peak, 8,408 feet.

Other major peaks: Rattlesnake Ridge, 7,748 feet; Massacre Peak,
    5,667 feet.

Dimensions: 6 x 20 miles.

Ecosystems: Chihuahuan Desert grasses, mesquite, piñon-juniper,
    Gambel and Emory oak, and mountain mahogany.

Counties: Luna.

Administration: Mostly BLM—Las Cruces Field Office, state, private.

Wilderness: The BLM Cookes Range Wilderness Study Area includes
    19,608 acres.

Getting there: From NM 26, at the locality labeled Florida on most maps, dirt County Road 19 runs northwest toward Cookes Peak, ending after 11 miles at a locked gate. From the junction of NM 26 and US 180 just north of Deming, County Road 8 runs north along the range's west side.

From earliest human history here, the stark fang of Cookes Peak has been the pivotal presence in southwestern New Mexico, anchoring this region as Sierra Blanca and Mount Taylor do theirs. The peak is an inescapable landmark, visible to travelers as distant as the Rio Grande, while Cookes Spring at the mountains' southeast base is one of only a few reliable water sources between the Rio Grande and the Mimbres River. More drama—usually tragedy—has played out around this mountain than perhaps any other summit in the state.

It was the spring, Cookes being the most recent of its many names, that first brought Native peoples here. Members of the Mogollon Culture who lived along the Mimbres River are called Mimbreños, and near Cookes Peak they built one of their classic four-room dwellings. A site near Cookes Spring has been tentatively identified as having been constructed by people of the Mimbreño Culture (AD 850–1350). Before the Mimbreños abandoned the area, for reasons still not fully understood, they created petroglyphs whose designs include masks, lizards, a plumed serpent, and birds.

The Mimbreños had vanished by the time Athabaskan peoples who call themselves N'de moved into the area, sometime after AD 1400. Known to Hispanics and Anglos as Apaches, they were a mountain people adapted to the desert. Raiding was part of their culture and their economy, and they preyed upon Mexicans, Americans, and other Indians. The vantage point of Cookes Peak, with its invaluable spring, was as precious to them as to other peoples.

Several Spanish military expeditions passed beneath the peak. The mapmaker-military officer Bernardo Miera y Pacheco labeled the peak *Cerro de los Remedios*, "peak of the remedies." In 1780 Captain Don Francisco Martínez, on an expedition coordinated by Governor Juan Bautista de Anza, camped by the spring, likely near the old Mimbreño settlement, and named it *San Miguel*. The present name, Cookes Peak, honors Captain Philip St. George Cooke, leader of the Mormon Battalion that passed through here in 1846–47.

It was the spring that led John Butterfield to establish a Butterfield

Cookes Peak

Overland Mail, or Pony Express, station at the spring and put the trail through narrow Cookes Canyon here, despite the extraordinary Apache threat along it. Soldiers at Fort Cummings, established in 1863 to protect travelers, called the gap at the head of Cookes Canyon a gauntlet of ambush sites, and gave no better odds than even of passengers making it through without an attack. Scores of people died here. In 1867 stage travelers complained that as many as nine skeletons were visible in Cookes Canyon; sensitive to public relations, the company removed them.

The Butterfield Overland Mail ceased in 1869, but westward travelers continued using the route. Fort Cummings closed in 1873, believing Apache attacks had ceased, but renewed raids led to its reopening in 1880. It closed finally in 1886, the year Geronimo surrendered. Today the fort is but a few deteriorating adobe and masonry walls surrounded by creosote bush.

But even as the Apaches were withdrawing, other people were invading the Cookes Range. In 1876 prospector George Orr found promising mineralization on Cookes Peak's eastern slopes. By 1882 a road was graded to the booming mining camp variously called Cookes and Cookes Peak. By 1911 the mining district had produced $3 million in lead, silver, zinc, and fluorite. The camp had a foul reputation. James A. "Jimmy" McKenna, author of *Black Range Tales* and a veteran of many mining camps, was justice of the peace at Cookes; he pronounced his clients "thieves, cutthroats, and outlaws," the worst he'd seen anywhere.

Today, the Cookes Range is tranquil. A few history buffs drive the dirt road to the meager ruins of Fort Cummings. The road branches from County Road 19, four miles from NM 26. By continuing on County Road 19 and leaving a vehicle at the locked gate of a private inholding, travelers can walk around the inholding and then hike the road for 2.7 miles to the weathered remains of Cookes. Despite the ubiquitous scatter of mine dumps, mine roads, stock tanks, and fences, the area is inexorably returning to its wildness; even now the solitude and silence are deep here. Except for occasional history buffs, rockhounds, and hunters, people seldom come here.

Certainly the odds of encountering hikers here are minuscule. There are no trails, and the obvious destination, Cookes Peak, is steep and difficult. A non-technical route to the top exists on the southeast side, avoiding the cliffs on the peak's other sides.

Cookes Peak so thoroughly dominates the Cookes Range that all the range's summits and ridges seem but spokes around a hub. The summit

Southern end of the Cookes Range, and Cookes Peak

spire rises abruptly, one thousand two hundred feet in less than a thousand feet above the mountain's approaches, themselves dauntingly steep. The peak is a huge, pointed cone of grandiorite, punched through Paleozoic and Cretaceous sedimentary layers about 38 million years ago. The range's southern part, low hills cut by deep canyons, exhibits mid- to late-Cenozoic basalt, andesite, and rhyolite ashflow tuffs.

Golden eagles nest on the cliffs of Cookes Peak, as well as hawks, owls, and falcons. Mule deer are common. A population of Gila whiptail lizards, previously known only from the Gila country, has been found here.

How ironic, that after so many centuries of human history, the history makers today are lowly lizards who've been here all along.

Note: Because of the extraordinary geological, ecological, archaeological, and historical values of the Cookes Range, as well as an increasingly wild natural environment, the NM Wilderness Alliance has recommended wilderness status for the wild lands here.

Cedar Mountain Range

Location: About 20 miles southwest of Deming, north of NM 9, east of Hachita.

Physiographic province: Basin and Range—Mexican Highland Section.

Elevation range: 4,700–6,200 feet.

High point: Flying W Mountain, 6,215 feet.

Other major peaks: Cedar Mountain, 6,207 feet.

Dimensions: 20 x 2.5 miles.

Ecosystems: Semi-desert grassland, chaparral, Chihuahuan Desert scrub
     including mesquite, creosote bush, and yucca, with scattered oaks.

Counties: Luna, Grant.

Administration: Mostly BLM—Las Cruces Field Office, state, private.

Wilderness: 14,911-acre BLM Cedar Mountains Wilderness Study Area.

Getting there: About 7 miles southeast of Hachita, a county road heads
     northeast from NM 9 into the mountains; other approaches require
     crossing private land.

The Cedar Mountains have no cedars, at least not true cedars. Actually, true cedars are found nowhere in New Mexico, the label misapplied usually to junipers, which like true cedars have red heartwood. To heighten the irony, few trees of any kind exist in the Cedar Mountains. Moreover, one could argue that use of the term "mountain range" itself is a bit overreaching, suggesting something far larger and higher. Still, compressed within the Cedars' long, narrow fetch is a daunting complex of canyons and ridges, with several very steep formations such as Hat Top Mountain (5,586 feet), Valiente Peak (5,765 feet), Turkey Knob (5,255 feet), and Deer Mountain (5,300 feet).

What trees exist in the Cedar Mountain Range are widely-spaced, low species such as juniper and hackberry. Rather, the vegetation is dominated by desert shrubs, including fourwing saltbush, Apache plume, tarbush, and mesquite. Though grazing exists in the Cedars, the mountains remain sufficiently wild that cattle share the habitat with quail, mule deer, pronghorns, javelinas, and raptors; hunting is the area's main recreational use.

Geologically, the Cedars are eroded expressions of the mid-Cenozoic volcanic rocks so common in southwestern New Mexico, along with small outcrops of Paleozoic and Mesozoic rocks at the north end. Connecting the high points along the slender, linear range is the Continental Divide, but poor access and respect for private land have caused the Continental Divide Trail to be routed elsewhere.

Florida Mountains in the early morning

Florida Mountains

Location: 10 miles southeast of Deming, east of NM 11.

Elevation range: 4,400–7,000 feet.

High point: Florida Peak, 7,448 feet.

Other major peaks: Gym Peak, 7,106; Baldy Peak, 6,980 feet; Capitol Dome, 5,962 feet.

Dimensions: 7.5 x 11 miles.

Ecosystems: Evergreen oak, juniper, bear grass, mesquite, ocotillo, cacti, and yucca.

Counties: Luna.

Administration: BLM—Las Cruces Field Office, state, private.

Wilderness: The BLM Florida Mountains Wilderness Study Area, 22,336 acres, includes the wild, generally inaccessible crest and related canyons and ridges.

Getting there: Though numerous roads lead into the Florida Mountains, the most popular access is from Spring Canyon State Park at the northeast end of the range. From Deming, take NM 11 south until NM 141 branches east; this becomes NM 143 leading to Rockhound State Park, but before the park, NM 198 branches, leading to Spring Canyon State Park.

A strand of barbed wire lying in the desert, or a linear patch of yucca—both images are evocative of the conspicuously jagged profile of the Florida Mountains. They remind some people of a huge battleship, perhaps afloat on one of the shimmering mirages so common here.

Yet like most fantasies, the Floridas shrink as one approaches. The range is small, and its tallest peak is not particularly high. It's the verticality of the peaks, rising some two thousand feet in less than two-thirds of a mile, along with the fantastic forms into which the mostly volcanic rocks have been eroded, that makes the Floridas so dramatic. Place names in the mountains reflect this: Castle Rock, Chimney Point, Lover's Leap Canyon, Needle's Eye, Devil's Arch, and Dragon Ridge.

No developed hiking trails exist in the Floridas, though two fairly clear tracks head into the backcountry from Spring Canyon State Park. Water sources also are few, mostly springs around the mountains' base. Indeed, about the only people who challenge the Floridas' ruggedness are either rockhounds or big-game trophy hunters out to bag a Persian ibex. In 1970, fifteen ibex were introduced into the Floridas because it was felt they would fill an unoccupied ecological niche in habitat that closely resembled that of the ibex's native Iran. The theory proved correct; the ibex have thrived. Now the concern is confining the ibex to the Floridas.

Such ranges as the Floridas are ecological islands, separated from neighboring Basin-and-Range mountains by expanses of inhospitable desert valleys. Thus, each range develops its unique species assemblage. The Florida array includes Gila monsters, javelinas, deer, mountain lions, coyotes, coatimundis, ringtails, and golden eagles, as well as other raptors nesting among the cliffs.

Geologically, the range is similar to others in the region. A core of Cambrian-Ordovician syenite and granite is overlain by Paleozoic and later sediments. Then during Cenozoic time, when all of the region was popping with volcanism, volcanoes here left a deep layer of volcanic breccia and other deposits. This activity resulted in relatively little mineralization, though miners from 1880 to 1956 extracted one hundred two thousand dollars in metals, mostly lead with small amounts of copper, silver, and gold, as well as fluorite and manganese.

Indeed, the richest gold deposits in these mountains are the yellow-orange poppies cloaking the foothills in the early spring, hence the Spanish

Venus, the Moon, and the skyline of the Florida Mountains

name *Florida,* "flowery." Even as early as 1762 Spanish speakers were referring to the range as the *Sierra Florida.*

In 1774 and in 1775, Spanish military expeditions noted that the Floridas
were Apache strongholds. They remained thus, and U.S. Army Buffalo
Soldiers fought at least two engagements in the mountains. On January 24,
1877, U.S. Army troops fought a battle with Apaches led by Victorio, whose
presence in the mountains is recalled by Victorio Canyon on the mountains' east side. Corporal Clinton Greaves earned a medal of honor when,
in the midst of hand-to-hand fighting, he forced his way through encir-
clement, enabling his comrades to break free.

Little Florida Mountains

Though the two neighboring Florida ranges share some geologic under-
pinnings and also a name, the Little Florida Mountains lack the rugged, dramatic profile of their larger sibling just half a mile to the southwest, across
Florida Gap. The Little Florida Mountains—stretching the term, for their
relief to a high elevation of about 5,250 feet is only about 1,000 feet—are an
elongate formation trending north-south about 2.5 x 6 miles.

The Little Florida Mountains' main significance lies in their geological
configuration. Composed of mid-Cenozoic volcanic breccia, rhyolite, and
tuff, the mountains from 1918 to 1959 were a major manganese producer
from mines on their northeast flank. The main ores were psilomelane, man-
ganite, and pyrolusite. Today visitors come to Rockhound State Park, on the
mountains' southwest side, to comb the rocky, treeless slopes for agates and
geodes. It is probably the nation's only park where visitors are encouraged
to pocket pieces of the natural environment as souvenirs!

Victorio Mountains

Location: 1.5 miles southwest of the Gage Exit on I-10, 18.3 miles west
    of Deming.
Physiographic province: Basin and Range—Mexican Highland
    Section.
Elevation range: 4,500–5,000 feet.
High point: Unnamed summit, 5,382 feet.
Other major peaks: None named.
Dimensions: 1 x 4.5 miles.

Ecosystems: Mesquite, creosote bush, and Chihuahuan Desert grasses
and shrubs.
Counties: Luna.
Administration: BLM—Las Cruces Field Office, state.
Getting there: From Gage, Exit 62 on I-10, County Road 20 heads south
2.25 miles, where several rougher roads branch west.

The Victorio Mountains are a cluster of steep-sided hills, aligned roughly
northwest-southeast. They are predominantly Lower Paleozoic limestone
and dolomite, overlaid by Cretaceous conglomerate, limestone, and sandstone. During the mid-Cenozoic upheavals in this part of the state the sedimentary rocks were intruded by mineralized granite and andesites. In 1880,
prospectors found silver deposits, and soon mines clustered around the
Chance and Jessie claims. Before long a little camp—sixteen wood-frame
buildings and two adobe structures—had sprung up. Initially it was called
Victorio, for the Apache leader, then Fullerton, when a post office was
established. But in 1886, and with characteristic miner optimism, the camp
became Chance City, for the Chance Mine. Most of the ore was argentifer-
ous galena, though zinc and gold also were produced. Total production
from the mining district has been estimated at $1.5 million. The last ore
shipment was in 1947. Now just rubbish and ruins remain, while the limestones are mined for aggregate for road construction.

Tres Hermanas Mountains

Location: Northwest of Columbus, west of NM 11.
Physiographic province: Basin and Range—Mexican Highland Section.
Elevation range: 4,400–5,400 feet.
High point: North Peak, 5,801 feet.
Other major peaks: Middle Peak, 5,786 feet; South Peak, 5,614 feet;
Black Top, 5,025 feet.
Ecosystems: Mesquite, tarbush, creosote bush, oak, mountain mahogany,
and manzanita.
Counties: Luna.
Administration: BLM—Las Cruces Field Office, state, and private.
Getting there: From Columbus, numerous jeep trails head into the
mountains; avoiding private land always is a concern.

Tres Hermanas Peaks

A tricorn of three steep and pointed peaks, all of comparable height, is the core of the "three sisters" complex. Though the group includes other peaks, it's the "sisters" that dominate the southern horizon from as far north as Deming. In the 1880s, the tale was told around Columbus that the "sisters" individually were named Alice, Kate, and Lou. Perhaps; those are just the sort of names miners and prospectors would bestow, and the Tres Hermanas Mountains participated in the general mining hubbub of south-western New Mexico in the late nineteenth century. Prospect pits are ubiquitous in the mountains, and by 1905 mining operations were underway, continuing until 1920 and sporadically since then. The main mines were on the northwest, mostly lead and zinc but with significant silver, gold, and copper. The mountains still attract prospectors, though now they are typically looking for mineral specimens rather than ores.

This is harsh, rugged, ragged desert country. Dry. The tiny settlement of Hermanas just to the west holds the state record for least annual rainfall: one inch, in 1910. The rocks here, readily exposed by mining and sparse vegetation, are the familiar, southwestern New Mexico, mid-Cenozoic volcanics and intrusions—andesite, latite, quartz monzonite—cutting much older Cretaceous and Paleozoic sedimentary formations. Major faults bound three sides of the range. Travertine and manganese oxide deposits on the northeast may reflect ancient hot springs, as the mountains are just west of the Las Cruces geothermal field. The group begins just south of the aptly named Greasewood Hills, 4,502 feet in elevation, then runs south about nine miles.

Sierra Rica

Location: At the instep of the Bootheel, south-southeast of Hachita, east of the Hachita Valley.
Physiographic province: Basin and Range—Mexican Highland Section.
Elevation range: 4,800–5,200 feet.
High point: Unnamed summit, 5,495 feet (in the United States); approximately 5,550 feet in Mexico.
Other major peaks: None named.
Dimensions: 6 x 12.5 miles.
Ecosystems: Chihuahuan Desert grasslands and shrubs, including evergreen oaks, mesquite, and creosote bush.
Counties: Hidalgo, Luna.
Administration: BLM—Las Cruces Field Office, state.

The *Sierra Rica* is a crescent-shaped group that begins on the north in New Mexico, runs south about two miles to enter Mexico, where it curves west to re-enter New Mexico south of the Apache Hills, an arc extending about 12.5 miles. Like the Little Hatchet Mountains to the west, the Sierra Rica consists of thick Paleozoic to Cretaceous strata, thrust to the northeast and intruded by mid-Cenozoic volcanics. And as in the Little Hatchets, this process resulted in mineralization, hence the name, "rich range," though the richness was more wishful thinking than descriptive. In the late 1800s, a small mining camp called Sierra Rica was here, and throughout the range are abandoned mines and prospects. Mines in the former Fremont Mining District produced minor amounts of silver, copper, gold, zinc, and lead.

Today, the range's dual-citizenship, poor access, harsh desert environment, and general obscurity make it an unlikely destination for hikers, especially as most of the range—and its highest point—is in Mexico.

Apache Hills

Location: 2 miles northwest of the Sierra Rica.

Physiographic province: Basin and Range—Mexican Highland Section.

Elevation range: 4,900–5,400 feet.

High point: Apache Peak, 5,760 feet.

Dimensions: 3 x 5.5 miles.

Ecosystems: Chihuahuan Desert grasslands and shrubs including evergreen oaks, mesquite, and creosote bush.

Counties: Hidalgo.

Administration: BLM—Las Cruces Field Office, state.

Getting there: From NM 81 and NM 9 south and east of Hachita, dirt roads lead toward the Apache Hills.

The Apache Hills geologically are a northwestern extension of the Sierra Rica and are similarly a complex of Paleozoic and later sedimentary rocks, intruded and altered by Cenozoic volcanics. In the process, copper, silver, bismuth, lead, zinc, and other minerals were deposited. Mines dating from the late 1800s, mostly on the mountains' southwest side, are now all abandoned.

Little Hatchet Mountains

Location: Southeast of Lordsburg, southwest of Hachita in the Bootheel, south of NM 9.

Physiographic province: Basin and Range—Mexican Highland Section.

Elevation range: 4,750–5,900 feet.

High point: Hachita Peak, 6,639 feet.

Other major peaks: Playas Peak, 5,863 feet.

Dimensions: 4 x 15 miles.

Ecosystems: Chihuahuan Desert grasses and shrubs including ocotillo, agave, yucca, and creosote bush.

Counties: Hidalgo, Grant.

Administration: BLM—Las Cruces Field Office, state.

Getting there: On the north, accessible from Hachita, is a rough dirt BLM road that parallels the mountains on the east; it leads to Hatchet Gap and NM 81 in the south.

Separated by Hatchet Gap from the Big Hatchet Mountains to the south, the

Little Hatchet Mountains are part of the same uplift that created the higher range, as well as the Brockman and Coyote Hills to the north, though the rocks in the Big Hatchets are older, mostly Paleozoic. Geologically, the Little Hatchet Mountains consist mostly of Lower Cretaceous sedimentary layers, complexly faulted and intruded by later volcanic rocks, a pattern repeated throughout southwestern New Mexico. Overthrust faults here led to overturned folds, further complicating the topography. In the Little Hatchet Mountains, this activity spawned extensive mineralization; some fifty-eight minerals have been recorded in Little Hatchet mining districts.

Prehistoric Indians, likely of the Jornada Mogollon Culture (AD 850–1350), mined turquoise at the range's north end, around Turquoise Mountain. The Indian workings were rediscovered in 1878, and soon the little mining camp of Hachita sprang up to house miners working claims not only for turquoise but also for silver, lead, and copper; by 1884 Hachita had three hundred residents. By 1890, however, mining had begun to decline. Then in 1900 the El Paso and Southwestern Railroad laid tracks nine miles north and east, siphoning the remaining residents to the community of New Hachita (now simply Hachita). Of Old Hachita, little remains today but desolate and long-abandoned buildings and mine debris. That's more than remains of the ephemeral mining camp of Sylvanite, located in Broken Jug Pass.

Except for Hachita Peak and Playas Peak, the range presents a profile of ridges more than defined summits. (Playas Peak is impressively stark and rugged when viewed from the northwest, on NM 9.) These are desert mountains; even their tops don't support trees. Instead, the vegetation is dominated by the classic plants of the Chihuahuan Desert—mesquite, creosote bush, ocotillo, mariola, several species of cacti and yucca, and enough grass to support the widely-dispersed cattle that graze here. Wildlife is limited to a few hardy mule deer and javelinas—and abundant quail, hunters report. A Continental Divide Trail (CDT) hiker reported seeing a coatimundi walking "in a sort of funny hunched-up run, not unlike the sea otters on the coast. . . ." (personal communication). Desert bighorn sheep were reintroduced successfully in 1979 and 1982.

Hiking is limited primarily to the Continental Divide Trail, which follows the rough BLM road paralleling the range along its eastern foothills. (The Continental Divide itself does not go along the Little Hatchet Mountains.) Hikers here confront exposure to wind and heat but especially paucity of water; approximately twelve miles separate windmill-fed stock tanks at the south and northern ends of the range. Local lore says springs once flowed in the mountains, but they have dried up with the pumping of ground water.

Actually, things weren't much better approximately 150 years ago. On Monday, November 23, 1846, the Mormon Battalion, recruited for the Mexican War and led to California by U.S. Army Captain Philip St. George Cooke, arrived at the Little Hatchet Mountains desperate for water. They found only a trickle where a spring had been hoped for, "only enough for ten men, not enough for the whole battalion. Men would try to lap up the water while on their bellies, or use their spoons to get precious drops from between the rocks."

Big Hatchet Mountains

Location: In the Bootheel, southeast of Lordsburg, southwest of Hachita, south of NM 81.

Physiographic province: Basin and Range—Mexican Highland Section.

Elevation range: 4,400–8,366 feet.

High point: Big Hatchet Peak, 8,320 feet.

Other major peaks: Zeller Peak, 7,313 feet, at the north end of the range.

Dimensions: 13 x 4.5 miles.

Ecosystems: Evergreen oak, piñon and Chihuahuan pines, semi-desert grassland, Chihuahuan Desert scrub.

Counties: Hidalgo.

Administration: Mostly BLM—Las Cruces Field Office, state, private.

Wilderness: The BLM Big Hatchet Mountains Wilderness Study Area includes 48,720 acres.

Getting there: Public access to the Big Hatchet Mountains is very poor, with most access blocked by private land. A rough dirt BLM road at Hatchet Gap heads southeast from NM 81 and parallels the range's eastern side before passing between the Big Hatchet Mountains and the Alamo Hueco Mountains.

When viewed from the east, from the Hachita Valley, the Big Hatchet Mountains are remarkably reminiscent of the Guadalupe Mountains at the opposite side of the state. Here as there, the mountains burst from a broad, featureless valley as an abrupt palisade of pale limestone cliffs atop steep, rocky foothills. Here as there, in an environment of almost no water, the sparse, harsh desert plants are little protection against the searing desert sun. Remote desert mountains, yet while most New Mexicans have at least heard of the Guadalupes, the Big Hatchets remain all but unknown in the instep of the Bootheel.

Geologically, the Big Hatchets, like the Little Hatchet Mountains, Coyote Hills, and Brockman Hills to the north, are expressions of a north-south uplift between the Hachita Valley on the east and the Playas Valley on the west. In the cliffs of the Big Hatchet Mountains, geologists read layers like pages in a book. Usually some pages are missing, but here the section is remarkably complete, beginning in the Precambrian. The sedimentary deposits here are mostly Paleozoic, much older than the deposits in the Little Hatchets. Big Hatchet Peak is an example of what geologists call an "over-thrust fault," where one layer overrides another, with the result that older rocks can be on top of younger ones, contrary to the normal sequence.

Probably the man who knew the Big Hatchets and their geology best was the geologist Robert A. Zeller, who lived in Hachita and wrote his Ph.D. dissertation about the mountains. After he died in a plane crash in 1970, a major peak at the range's north end was named for him. The range's highest peak—Big Hatchet Peak—already was named and leading to its steep-sided summit is the range's only defined hiking trail, a faint, unmarked track leading from a windmill up Thompson Canyon on the range's east side.

The Big Hatchets aren't exactly a magnet for hikers, though occasionally hikers come here, well . . . just to see what's here. A few hikers come here to begin their trek on the Continental Divide Trail, for while the Divide itself doesn't go through the mountains, the CDT does. So recently established is this, however, that no visible tread exists on the ground, and the CDT route is just that, and subject to change. All hikers in the Big Hatchets confront infrequent to non-existent surface water; for extended trips hikers must depend upon widely-spaced windmills and stock tanks, marked on maps but not always maintained or operating. Lacking trails, hikers must go cross-country and often must thrash through thickets of mesquite, cholla, prickly pear cactus, ocotillo, and the jackstraw-strewn stalks of century-plant agaves. Tough going. But it's also a place of unexpected beauty, and following a wet winter the mountains are a miracle of desert wildflowers. Rare or endangered plants here include nightblooming cereus and Scheer's pincushion cactus.

And for all their aridity and ruggedness, the mountains are home to a surprising diversity of animals, including one of the Southwest's few remaining indigenous herds of desert bighorn sheep. To support them the BLM has constructed cisterns to catch rainfall. Among other rare species here are the Sonora mountain king snake, thick-billed kingbird, varied bunting, coatimundi, and giant spotted whiptail lizard. The Vallonia snail (*Vallonia sonorana*) and the fringed mountain snail (*Radiocentrum ferrissi*) are found only in the Big Hatchets.

Apaches once roamed the Big Hatchets. Preceding them were Indians of the Mogollon Culture (AD 850–1350); their artifacts have been found in the range's caves. Unlike the Little Hatchet Mountains to the north, the Big Hatchets have seen little mining or prospecting, primarily because of remoteness and paucity of valuable minerals, though galena, gypsum, malachite, smithsonite, and sphalerite are found here.

Alamo Hueco Mountains

Location: In the extreme southeastern corner of the Bootheel.
Physiographic province: Basin and Range—Mexican Highland Section.
Elevation range: 4,800–6,800 feet.
High point: Unnamed summit, 6,838 feet.
Other major peaks: Pierce Peak, 6,149 feet.
Dimensions: 5 x 9.25 miles.
Ecosystems: Evergreen oak, piñon and Chihuahuan pines, Chihuahuan
    Desert scrub and semi-desert grassland.
Counties: Hidalgo.
Administration: Mostly BLM—Las Cruces Field Office, some private.
Wilderness: The BLM Alamo Hueco Mountains Wilderness Study Area
    is 16,264 acres.
Getting there: Public access to the Alamo Hueco Mountains is
    extremely poor, not only because of remoteness but also because
    the mountains are almost entirely surrounded by private land.
    A rough dirt BLM road heads southeast from Hatchet Gap on NM 81,
    and after paralleling the Big Hatchet Mountains' eastern side, heads
    west to pass between the Big Hatchets and the Alamo Huecos.

Viewed from the north, especially at sunset, the Alamo Huecos seem a landscape of bare, orange-red knobs and protuberances surreal in their shapes and colors. Mountains of mystery and beauty, an apt frame to put around a range that despite its size is little known to the public.

The Alamo Huecos occupy a remote cranny in the vast, empty Bootheel, bordered on two sides by Mexico, surrounded by private land as with barbed wire; these are the original "can't get there from here" mountains. And sure enough, few people do, even during the brief period of less-than-torrid weather. Until recently, when the BLM opened a long, rough road to the north, the public had no legal access to its lands here. Indeed, many New Mexico outdoor guide books don't even include the Alamo Huecos, or they subsume them under the larger Big Hatchets to the north, despite the Alamo Huecos being physically distinct.

Humans haven't always been absent from the Alamo Huecos. In remote caves the archaeologists Harriet and Cornelius Cosgrove in 1947 found evidence of Mimbres Mogollon habitation, circa AD 1000. More recent habitation is evinced by the weathering remains of long-abandoned homesteads: a tiny ranching locality named Alamo Hueco is in the mountains' western foothills. In 1923, in perhaps the last incident in the Apache conflicts, a posse trailed an Apache band heading south past the Alamo Huecos. The Apaches had been living in Mexico and had crossed into the United States to steal livestock in the Bootheel. The stolen animals were recovered, but the Apaches made it back to Mexico. The mountains continue to have renegade associations: in the 1990s, the Alamo Huecos were along a smuggling route used by the Aguirre drug running operation out of Deming.

Ecologically as well as culturally, the range is a frontier, and it has remained wilderness. Most of the wild land is within the Mexican Highland Shrub Steppe ecological zone, yet the mountains' habitats are surprisingly diverse, as are the animals in them: javelinas, mountain lions, golden eagles, Montezuma quail, and mule deer. Desert bighorn sheep, long absent, were reintroduced in the fall of 1986. Rare and endangered species in the general area include Sonora mountain kingsnake, thick-billed kingbird, varied bunting, coatimundi, giant whiptail, and jaguar. Rare and endangered plants include nightblooming cereus and Scheer's pincushion cactus.

Within the Alamo Huecos' approximately thirty thousand acres, only one summit is named: Pierce Peak. The Spanish name *Alamo Hueco* itself means "hollow cottonwood," and indeed Cottonwood Spring, draining into

Cottonwood Canyon, is among the range's few named features. In these dry mountains, any cottonwood, hollow or otherwise, would be notable.

Geologically, the Alamo Huecos are mostly made up of ashflow sheets, 35.1–27.4 million years old, tilted into fault blocks, with small Cretaceous outliers. Basaltic flows are on top and down the southwest side.

No hiking trails exist in the Alamo Huecos, though a few very rough two-tracks probe the area. All hikers here must address serious issues of heat, rough and confusing topography, and the paucity of water, as well as the possibility that hikers will be received less than warmly by local people and border law enforcement.

Dog Mountains

Straddling the United States-Mexican border, about 4.6 miles long with 2.9 miles in New Mexico, the Dog Mountains are an appendage of the Alamo Hueco Mountains and are separated from them on the north by Horse Canyon. The Dog Mountains share the geology and ecology of the Alamo Huecos, as well as their abrupt, steep profiles. The highest point, in the northeast part of the range, is an unnamed summit, 6,552 feet. No trails exist, though several dirt roads penetrate the area. While the mountains themselves are mostly public land, they're surrounded by private land.

As for the name, its origin may not be as simple as it seems, for as *The Place Names of New Mexico* explains, the name may not refer to domestic canines, "or even to coyotes but rather to prairie dogs."

Pyramid Mountains

Location: 2 miles south of I-10 at Lordsburg, east of NM 338, on the east
   side of the northern Animas Valley, north of NM 9.
Physiographic province: Basin and Range—Mexican Highland Section.
Elevation range: 4,580–5,400 feet.
High point: Pyramid Peak, 6,008 feet.
Other major peaks: South Pyramid Peak, 5,910 feet; Rimrock Mountain,
   5,779 feet; Aberdeen Peak, 5,048 feet; Lee Peak, 5,012.
Dimensions: 6 x 20 miles.
Ecosystems: Oak, juniper, mountain mahogany, manzanita, saltbush,
   yucca, and cacti.
Counties: Hidalgo.

Administration: BLM—Las Cruces Field Office, state, private.

Getting there: From Lordsburg, NM 494 and County Road 9 head southeast. Farther south, several dirt BLM roads head into the mountains.

It's hard to understand today the importance these desiccated, ripsaw-
rugged, desert peaks once had in the region's history. Pyramid Peak, certainly, would have been a landmark for any early travelers; it rises one
thousand two hundred feet steeply on all sides, with a compelling resemblance to a pyramid.

In 1852, Eugene Leitendorf dug wells west of Pyramid Peak as he was
driving cattle from Illinois to California. Leitendorf's wells became a water
stop on the overland route to California, and his name became attached to
the little settlement that sprang up there. The Leitendorf Hills, just west of
Pyramid Peak across Gore Canyon, are a steep, compact cluster, highest elevation 5,349 feet.

Just east of the settlement of Leitendorf, at the base of Pyramid Peak,
was the vigorous but transitory mining camp of Pyramid, which with ironic
optimism also had been called Pyramid City. The town was born with mineral discoveries in the early 1880s, and by 1882 the town had a post office. It
was part of the same mining frenzy that had created the communities of
Valedon and Shakespeare at the Pyramid Mountains' northern end. Copper
and silver were the main ores in the Pyramid Mining District, with smaller
amounts of lead, zinc, and gold. Between 1870 and 1973 the district produced more than $59 million in these minerals.

But not diamonds. In the 1870s, some miners in the nearby camp of
Ralston hoped to lure investors by reporting a diamond strike on Lee Peak.
The scam was exposed, and the town of Ralston, eager for a new identity,
renamed itself Shakespeare. At Pyramid, the mining enthusiasm soon
fizzled; in 1897 the village's post office closed.

The Pyramid Mountains owe their topography—as well as their minerals—to two periods of volcanism. The first was associated with the
Laramide Orogeny and began around 70 million years ago. This resulted in
some volcanic rocks and intrusions in the range's northern part. The second produced andesite flows and other volcanics from a caldera 35.2 million years ago. Eruptions and intrusions along a north-south line of
weakness produced the linear range, characterized by a few high-relief
summits surrounded by much more modest hills. Pyramid Peak and South

Pyramid Peak are volcanic plugs. Like the Animas Mountains to the south, the Pyramid Mountains separate the Playas Valley on the east from the Animas Valley on the west—classic Basin and Range topography. The Continental Divide winds through the mountains, reaching South Pyramid Peak before abruptly heading east.

Animas Mountains

Location: North-south through the center of New Mexico's Bootheel, east of County Road 1.

Physiographic province: Basin and Range—Mexican Highland Section.

Elevation range: 4,750–8,500 feet.

High point: Animas Peak, 8,519 feet.

Other major peaks: Center Peak, 7,062 feet; Gillespie Mountain, 7,324 feet; Rough Mountain, 6,612 feet.

Dimensions: 40 x 12 miles.

Ecosystems: Chihuahuan Desert scrub and grasses, piñon and alligator juniper, Madrean pine-oak, mixed conifer.

Counties: Hidalgo.

Administration: Mostly private at the range's southern end, BLM—Las Cruces Field Office in the north.

Getting there: The portions of the Animas Mountains administered by the Animas Foundation for ecological research are closed to the public. Vehicular access would be from County Road 1 in the Animas Valley. At the range's south end, the road over San Luis Pass, closed to the public, goes between the Animas and San Luis Mountains.

One cannot escape the suspicion that if the sprawling Animas Mountains—running approximately forty miles south from NM 9 to San Luis Pass, as much as twelve miles wide—were in any region other than New Mexico's remote Bootheel, this mountain complex would be balkanized geographically into several ranges, each with its own name and identity. Instead, the land here is too people-empty to indulge such distinctions. What unites all the mountains is that they lie between the Playas Valley on the east and the Animas Valley on the west. And all are strung together along the Continental Divide, like beads on a string. Geologists tell us that once these mountains overlooked huge Pleistocene lakes in the Animas and Playas Valleys, as well as Lake Cloverdale.

In this configuration, the Animas Mountains epitomize the Basin and Range Province: a linear north-south range, uplifted between two subsiding basins. In the range's northern part, these hemispheric-scale forces have cracked and faulted the rocks, creating the range's present complexity. Concommitant with stretching was volcanic activity, and much of the mountain topography here results from three volcanoes that formed calderas here 35.1, 33.5, and 33.5 million years ago. These calderas and stacks of ashflow deposits have been broken and tilted into fault blocks. Cretaceous and even older rocks, dating to the Paleozoic and Precambrian Eras, are exposed at the north end and often are manifested as low, angular ridges fringed with erosional pediments.

Farther south in the range, the older rocks are overlain by mid-Cenozoic volcanics. Despite this geologic activity, mineral prospectors have never found much of interest in the Animas Mountains, though one mine is recorded as having produced some silver.

To southwestern ecologists, however, the Animas Mountains are exceptionally rich. The range has been estimated to include seven hundred plant species, seventy-five animal species and subspecies (including the rare, yellow-nosed cotton rat and the hog-nosed skunk), and more than fifty amphibian and reptile species, such as the New Mexico ridgenose rattlesnake (*Crotalus willardi obscurus*), found only here and in the nearby San Luis Mountains. Entomologists also find rare and undescribed species in the Animas Mountains. Located between the Chihuahuan and Sonoran Deserts and rising in layers over half a mile from the adjacent basins, the life zones here epitomize the Sky Island concept, where isolated mountains rise above desert basins, separated from other mountain ecosystems like ocean islands in an archipelago. Thus, each range has a distinct mix of species. In New Mexico, the madrean pine-oak habitat is found only in the Animas Mountains and in the nearby Peloncillo Mountains.

Because of their isolation from humans, the Animas Mountains have been mountains of refuge, albeit imperfect, for numerous animals. In the rugged Animas Mountains were New Mexico's last free-roaming Mexican gray wolves; the last was killed here in 1938, and unconfirmed sightings continued into the 1940s. By the late 1800s, desert bighorn sheep had become extinct here, though horns and sightings were reported as late as 1893. In 1981, nearly a hundred years later, sightings were reported again. Around 1911, the famous (or infamous) professional hunter Ben Lilly worked in the

Animas Mountains; he killed thirteen lions, several grizzlies, twelve black bears, and a few wolves. Of those species, only lions and black bears remain. Now ecologists rather than hunters track the animals.

The focus for most ecological research here is the former Gray Ranch, a three hundred twenty one thousand-acre parcel once part of the historic Diamond A Ranch. In an ecological coup, the Nature Conservancy in 1990 acquired the property; it was at the time the world's largest nature conservation acquisition, five hundred square miles that included much of the Animas Mountains. The property is managed by the Animas Foundation for long-term ecological studies, including the effects of fire and grazing, mesquite ecology, montane rattlesnake research, and monitoring of such rare species as white-sided jackrabbit, Botteri's sparrow, Baird's sparrow, Mexican chickadee, yellow-eyed junco, grasshopper sparrow, Sprague's pipit, Chiricahua leopard frog, bunchgrass lizard, and Gould's turkey. The elegant trogon, a Mexican species much sought by birdwatchers, sometimes is encountered here.

The Animas Foundation restricts public access to its Animas Mountains holdings, which is why the Continental Divide Trail doesn't go along the Animas Crest, as the Continental Divide itself does. This also makes the Animas Mountains one of four major mountain ranges in New Mexico closed to the public, the others being the San Andres and Oscura ranges within White Sands Missile Range, and the Fra Cristobal Range, part of Ted Turner's holdings. For these and other reasons, hiking opportunities here are extremely limited.

Historically, human presence here has been minimal. The terrain is daunting—steep and rocky—the climate searing, and water sparse. In cooler, wetter periods, Paleo-Indians would have hunted large game and gathered plants here. They were succeeded around seven thousand five hundred years ago by other nomads, of the Archaic Culture, that also came, camped, and departed, leaving little evidence of their presence. By seven hundred years ago, Archaic hunter-gatherers had become agriculturists. Indians of the Mogollon Culture lived at the base of the Animas Mountains in a pueblo that included a ball court, linking it with Casas Grandes cultures to the south in Mexico.

In historic times, the Animas Mountains were controlled by Apaches, who blocked most settlement until the Apache wars ended in the 1880s. (In 1877 U.S. Army Sixth Cavalry troops reported killing fifteen Apaches in a battle in the Animas Mountains.) The name *Animas* perhaps is linked to this violent era. It means "souls," and elsewhere in the Southwest abbreviates the phrase *Las Animas Perdidas,* "the lost souls," likely referring to Catholics who died without final rites, as in a frontier skirmish. It's unknown whether the name was applied first to the mountains or the Animas Valley.

Even now, the Bootheel accommodates only a few, widely-spaced ranches—and no settlements. If you ever decide to visit Cloverdale, which appears as a locality on some maps, you'll be sorely disappointed. The sole building is but a shelter for cows.

Whitewater Mountains and Smuggler Hills

Approximately 6 miles southeast of the southern end of the Animas Mountains and related to them are the Whitewater Mountains, a compact, 3.7 x 2.5-mile knot of steep summits on the United States-Mexican border. None are named, at least on USGS 7.5-minute quadrangles; the highest is 5,887 feet in elevation.

Just to the north of the Whitewater Mountains and separated from them by Deer Creek are the Smuggler Hills, a narrow, southeast-trending appendage of faulted and tilted volcanic rocks, 9 miles long by 1.5 miles wide, of the Animas Mountains. The "hills" actually are quite steep and rugged; they are capped in the south by 5,995-foot Hilo Peak.

Even more closely related to the Animas Mountains are the San Luis Mountains (see below).

San Luis Mountains

Location: South of the Animas Mountains, extending into Mexico.
Physiographic province: Basin and Range—Mexican Highland Section.
Elevation range: 5,200–6,000 feet, in New Mexico.
High point (in New Mexico): Unnamed point, 6,787 feet.
Dimensions (in New Mexico): 2.5 x 4.5 miles.
Other major peaks: None.
Ecosystems: Chihuahuan Desert scrub and grasses, including piñon
    pines and alligator juniper, and Madrean oak.
Counties: Hidalgo.
Administration: Private and BLM.
Getting there: The dirt road connecting the Animas Valley and the

Playas Valley via 5,522-foot San Luis Pass is the best access, though it is presently closed to the public. It passes the Animas Mountains to the north and the San Luis Mountains to the south.

The San Luis Mountains appear on New Mexico maps as just an appendage to the adjacent Animas Mountains, but they actually are the northern extension of a far larger, far rougher range in Mexico, known there as the Sierra San Luis. Formed 36–25 million years ago of the same mid-Cenozoic volcanic rocks as the Animas Mountains, the San Luis Mountains also share with them an interesting array of Chihuahuan Desert plants and animals, many at the northern extent of their ranges. The white-sided jackrabbit, rare in the United States, has been reported in the San Luis Mountains' eastern foothills.

The San Luis Mountains are separated from the Animas Mountains by San Luis Pass. Geologically and ecologically, the two ranges are one (see *Animas Mountains*). It is atop the crest of the San Luis Mountains that the Continental Divide begins its long journey northward in the United States. (Because of land-ownership issues, the Continental Divide Trail begins farther east.)

Peloncillo Mountains

Location: Begin in extreme southwestern New Mexico and run north along the Arizona border, crossing it, then extending north in Arizona to the Gila River.

Physiographic province: Basin and Range—Mexican Highland Section.

Elevation range: 4,000–6,000 feet.

High point: Grey Mountain, 6,928 feet.

Other major peaks: Black Point, 6,467 feet; Guadalupe Mountain, 6,444 feet.

Ecosystems: Chihuahuan Desert trees and shrubs including mesquite, yucca, agaves, cacti, bear grass, alligator juniper, mountain mahogany, piñon, live oaks, and Chihuahuan pines.

Counties: Hidalgo.

Administration: Coronado National Forest—Douglas, Arizona, Ranger District; BLM—Las Cruces Field Office; private.

Wilderness: Whitmire Canyon Wilderness Study Area, 18,000 acres.

In Arizona, BLM Peloncillo Mountain Wilderness (Safford, Arizona

BLM Office), 19,440 acres, 9 miles northeast of San Simon; Coronado National Forest—Douglas, Arizona.

Getting there: From Exit 5 on I-10, NM 80 heads south then southwest to pass through Granite Gap at about 10 miles. Also from I-10, at Exit 11, NM 338 heads south to Animas, from whence NM 9 crosses the range to the west. Pavement on NM 338 continues south of Animas, but after several miles the pavement ends and maintained dirt begins. From this road, about 29 miles south of Animas, Forest Road 63 heads west along Clanton Draw, with a pleasant intermittent stream. It begins its descent from the high point 1.5 miles before entering Arizona.

In 1996 the Arizona rancher and hunting guide Warner Glenn, while scouting lions in the New Mexico Peloncillo Mountains for a client, came upon extraordinarily large cat tracks. Following the creature, Glenn finally confronted—a jaguar. To his eternal credit, Glenn didn't shoot the animal but instead took photos. In the 460 years since Coronado became the first European in New Mexico and reported *tigres* and *ounces* (*once* being a common Latin American word for jaguar), jaguars had been reported in New Mexico only twenty times—and of those only three presented hard evidence. Glenn's photos demonstrated that the Peloncillos are still within the jaguar's range. (See also *Black Range* and *Burro Mountains*.)

Biologists were not surprised by the location. Indeed, one biologist has said the Peloncillos are home to more vertebrate species than any U.S. national park. This ecological richness results from the mountains straddling the life zones of the Chihuahuan and Sonoran Deserts, as well as being a transition between the Sierra Madre ecosystems south in Mexico and more temperate habitats north in the United States. In addition to having typical mountain fauna (deer, coyotes, black bears, lions, foxes, bobcats, etc.) the Peloncillos also accommodate desert bighorn sheep (reintroduced in 1981), Coues deer (a western subspecies of white-tailed deer), javelinas, ringtails, coatimundis, Gila monsters, peregrine falcons, trogons, rare buff-collared nightjars, and many more. The last wild Mexican wolf in New Mexico was killed in the nearby Animas Mountains in 1938, but rumors persist that a few *lobos* may still haunt the Peloncillos; a researcher using baited camera stations to photograph coatimundis also photographed a large, wolflike canid. Botanists have said that in Granite Gap is New Mexico's

Granite Peak (left), Peloncillo Mountains

greatest cacti array. Even a short hike through this prickly desert garden lends credence to the claim.

It is for their ecological value that the Peloncillos are a critical component of the Sky Island Alliance's goal of creating a "super-wilderness" in southwestern New Mexico and southeastern Arizona. By linking ecological and administrative units, such as the Gila Wilderness, several Wilderness Study Areas, riparian areas, and others, the Alliance hopes to create corridors that will allow native species access to much more habitat than when stranded on their "sky islands" or in postage-stamp-size Wilderness Study Areas.

If the natural history here is dramatic, so is the human history. It's ironic that so many significant, or at least colorful, events occurred in this range now widely ignored. In 1846 U.S. Army Captain Philip St. George Cooke led the Mormon Battalion over the pass separating Clanton Draw in New Mexico from Cottonwood Creek in Arizona. In addition to confronting grizzlies, they had to use ropes to lower wagons and horses down the steep slopes.

In 1857 John Butterfield routed his Butterfield Overland Mail through the Peloncillos, crossing them at a place called Doubtful Canyon, named because the chances of not being ambushed by Apaches were doubtful. The previous year U.S. Army Captain Enoch Steen commanded U.S. Army troops who accompanied travelers through the gap, and his name, albeit misspelled, became attached to 5,867-foot Stein's Peak. When Butterfield established a Pony Express station here, he called it Stein's Peak Station. The Butterfield Overland Mail lasted only until 1869. When the Santa Fe-Pacific put a railroad line through the Peloncillos around 1880, they put it eight miles south of the old stage route, and a new Steins station was born. Though long decommissioned, the locale retains a few inhabitants.

Apaches had no monopoly on ambushes in the Peloncillos. In 1882 Curly Bill Brocius and his outlaw gang ambushed a pack train of Mexicans returning to Mexico, some accounts say with plunder from raids in Arizona. The outlaws killed fifteen Mexicans and left their bodies to animals and the elements. For years souvenir hunters with ghoulish tastes scavenged bones in what came to be called Skeleton Canyon. Other outlaws also operated out of the Peloncillos. The Clanton brothers, outlaw icons of Tombstone infamy, had a base in the mountains' eastern foothills and frequently used the route now bearing their name, Clanton Draw.

Then in 1886 an event far more significant occurred: in Skeleton Canyon 1.25 miles west of the New Mexico border in Arizona, Geronimo surrendered to the U.S. Army. On September 3, a renegade band led by Geronimo and Naiche—nineteen men and twenty-eight women and children—realizing that further flight was futile, turned themselves over to General Nelson A. Miles, who had been pursuing them guided by Apache scouts. With the surrender, almost three and a half centuries of armed conflict between Native Americans and Europeans, which began in 1539 when Spanish explorers under Coronado clashed with Zuni Indians, ended. The outcome, in the first battle as in the last, was the same: faced with overwhelming non-Native resources, the Indians ultimately had no choice but to accept the inevitability of capitulation. A new era in Native American-European relations began.

Arriving in the Peloncillos soon after the departure of Apaches were miners and prospectors. They found minor deposits of gold and tungsten, as well as fluorspar, lead, silver, copper, and zinc. The remains of mining are evident at Granite Gap, where seven hundred fifty thousand dollars in silver and lead were extracted. The Carbonate Hill Mine north of the gap produced $1.5 million. But by 1920 mining had all but ceased.

The Peloncillo Mountains' turbulent human history, with its quiescent present, parallels their geologic history. A geologist described the range as a typical complex rotated fault-block range of the Mexican Highlands. In the central part of the range, between Granite Peak and Steins Pass, are faulted Precambrian, Paleozoic, and Cretaceous rocks. Mid-Cenozoic volcanism also affected the range, which now exhibits the eroded remains of two calderas: the Steins Caldera, dating from 34.4 million years ago, and the Portal Caldera, 27.6 million years old. Young basaltic lava flows flank the range on both the New Mexico and Arizona sides. Two other calderas are south of the Peloncillos and contributed ashflow tuffs to parts of the range.

These volcanic events created mountains that often are less a continuous crest than a linear array of steep-sided volcanic mounds and peaks. The range tapers to low hills at its northern and southern ends, with the center around Steins having the highest, steepest peaks, often isolated from one another. Most I-10 travelers have no perception of passing through a mountain range. The Peloncillos begin at the Rio Guadalupe, just north of the Guadalupe Mountains at the bottom of the Bootheel, and run north along the Arizona-New Mexico border, between the Animas and San Simon Valleys. They leave New Mexico about 5 miles north of I-10 and continue north to end at the Gila River—a total reach of about 114 miles, about fifty-eight percent in New Mexico.

The pattern of distinct, often conical mountains likely accounts for the range's name. According to *The Place Names of New Mexico:*

> Two equally plausible explanations exist for this name. *Peloncillo* in Spanish means "little baldy," and that accurately describes the range's barren summits, particularly in contrast to the higher, forested Chiricahua Mountains to the west in Arizona. But peloncillo also could be a misspelling of *piloncillo,* which in the Southwest referred to conical pieces of brown, unrefined sugar—a "sugarloaf"—and numerous peaks in the range do indeed resemble this.

Guadalupe Mountains

Location: Extreme southwestern New Mexico, bordering Arizona and Mexico, west of the Animas Valley.

Physiographic province: Basin and Range—Mexican Highland Section.

Elevation range: 4,600–6,000 feet.

High point (in New Mexico): Guadalupe Mountain, 6,444 feet.

Other major peaks: Bunk Robinson Peak, 6,241 feet; Kilmer Peak, 6,088 feet.

Dimensions: See below.

Ecosystems: Piñon, alligator juniper, and Emory and Arizona white oak.

Counties: Hidalgo.

Administration: Private, BLM—Las Cruces Field Office; and Coronado National Forest—Douglas, Arizona, Ranger District.

Wilderness: The 7,000-acre Bunk Robinson Wilderness Study Area is here.

Getting there: Access to this extremely remote area is via dirt County Road 1, leading to Forest Road 63, which runs along Cloverdale Creek into Arizona. The Guadalupe Mountains are south of this road.

The Guadalupe Mountains are much larger than they appear on most New Mexico maps, for they extend south into Mexico and west into Arizona. At their northeast end, they are separated from the Peloncillo Mountains by the valley of Cloverdale Creek; they extend south about twelve miles to the Mexican border, a dense knot of ridges and canyons. Guadalupe Canyon runs southwest—northeast through the range. Most of the range is made up of mid-Cenozoic volcanic rocks related to the eruption of the Geronimo Trail caldera (32.7 million years ago) and the Clanton Draw caldera, 27.4 million years ago.

Like the mountains, the plants and animals here ignore political borders and thus species from several ecosystems mingle (see *Peloncillo Mountains*). Few hikers—or anyone else—ever come here, which suits the local people just fine.

CHAPTER SIX

Datil-Mogollon Section of the Transition Zone

Transitional between the Colorado Plateau Province to the north and the Basin and Range Province to the east and south, the Datil-Mogollon Section is dominated by the Datil-Mogollon Volcanic Field, where widespread volcanism created almost all of the region's many mountains: the San Mateo and Magdalena Mountains west of the Rio Grande Rift, as well as the Mogollon Mountains, the Black Range, and all the small ranges around the Plains of San Agustin. Reflecting their common origin, almost all of the mountains here exhibit a striking similarity in their rocks: light-colored, fine-grained volcanics rather than lavas or basalts. Also reflecting a common volcanic origin, the mountains of this province typically are dry, with little surface water on the highly porous volcanic rocks and soil. Nonetheless, several important drainages originate here, including the Gila, San Francisco, and Mimbres Rivers.

The largest and third largest wilderness areas in New Mexico, The Gila and Aldo Leopold, are here, as well as several other significant wildernesses, including the Withington, Apache Kid, and Blue Range Wildernesses. Also here are several large and important Wilderness Study Areas, including Horse Mountain and Continental Divide. But even mountains such as the Magdalenas, Datils, and Tularosas, where no formal wilderness designation has been proposed, nonetheless are conspicuously wild and natural, with human activity as intermittent as the streams. Within this vast and sparsely populated region are habitats ranging from desert and riparian to high-elevation, moist-temperate conifers. Accentuating the ecological diversity here is the fact that the province is within the transition between southern and northern life zones. Whitewater Baldy, 10,892 feet in elevation, is the province's highest point.

Datil-Mogollon Section of the Transition Zone

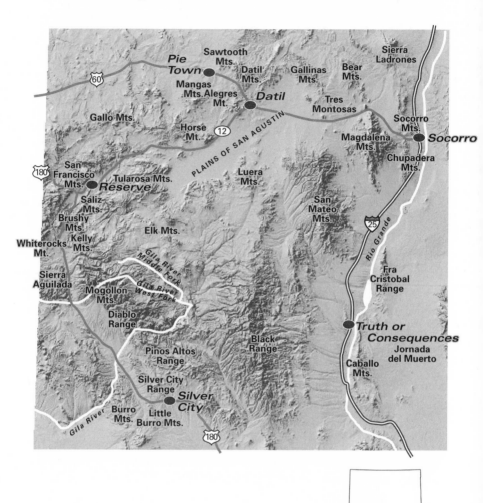

San Agustin Graben

A glance at a relief map of southwestern New Mexico shows many mountain groups located around a great enclosed basin, the Plains of San Agustin. The Plains are all that remains of a Pleistocene lake that measured 15 x 45 miles, was three hundred feet deep at the southwest end, and once saw Paleo-Indians camped on its shores. They hunted now-extinct mammals, as well as deer and elk, just as humans do today. Their descendants later experimented with agriculture; corncobs found in nearby Bat Cave are among the oldest in the Southwest.

To geologists, the Plains of San Agustin are the surface expression of a *graben*, a block of Earth's crust that slumps downward while adjacent areas are uplifted. This occurred here following the general volcanic upheavals in this part of New Mexico beginning about 36 million years ago. Surrounding the Plains, like volcanic islands in an atoll, are at least ten volcanic mountain formations. Clockwise from the north they are the Gallinas Mountains, Magdalena Mountains, San Mateo Mountains, Luera Mountains, Pelona Mountain, Tularosa Mountains, Mangas Mountains, Horse Mountain, Crosby Mountains, and Datil Mountains.

All are remote, seldom visited, generally lacking surface water, and vegetated with piñon-juniper and ponderosa pine forests, and high, open grasslands. The topography typically is complex, a net of steep ridges and deep canyons. Except for the Continental Divide Trail, which follows the Divide over Pelona Mountain and through the Tularosa and Mangas Mountains usually on Forest Roads, few hiking trails exist, though hiking on dirt roads or cross-country is relatively straightforward.

Bear Mountains

Location: North of the village of Magdalena, east of NM 52, south of the Rio Salado.

Physiographic province: Datil-Mogollon Section of the Transition Zone.

Elevation range: 6,400–8,000 feet.

High point: Unnamed summit, 8,221 feet.

Other major peaks: No named summits.

Dimensions: 4.5 x 14 miles.

Ecosystems: Piñon-juniper, Gambel oak, and mountain mahogany.

Counties: Socorro.

Administration: Cibola National Forest—Magdalena Ranger District.
Indian lands: The Alamo Chapter of the Navajo Nation is immediately
   northwest.

It's tempting to declare the name of these mountains a mistake. Surely the
namers meant "Bare," not "Bear," because there's so little vegetation here
that any bear would probably just be passing through. Actually, the name,
like many, probably refers to a specific incident involving a bear, not a concentration of them. And most likely the name was first applied not to the
mountains but to the major spring at the range's south end. Surprisingly,
springs are relatively common in the mountains, and it's these oases that
have given the Bear Mountains a reputation among birders. They, along
with a few geologists, are about the only non-local people to visit the Bear
Mountains.

   Geologically, the southern Bear Mountains are of a piece with the
Magdalena Mountains to the south (see entry), both aligned along the
Magdalena Fault. Most of the range consists of domino-tilted layers of ash-
flow tuffs expressed as cuestas, or long, linear ridges.

   To reach the range's high point, go north on Forest Road 354 then head
west on Forest Road 354A, paralleling Cedar Springs Canyon. After 4.4 miles
you reach Cedar Springs; at 5.4 miles you reach some prospects. And here
you confront the steep, eastern escarpment of the range and will have to
sight-read your way up the unnamed summit. Keep your eyes out for birds,
not bears.

Magdalena Mountains

Location: South and west of US 60, south of the village of Magdalena.
Physiographic province: Datil-Mogollon Section of the Transition Zone.
Elevation range: 5,800–10,400 feet.
High point: South Baldy, 10,783 feet.
Other major peaks: North Baldy, 9,858 feet; Magdalena Peak, 8,152 feet.
Dimensions: 15 x 25 miles.
Ecosystems: Gambel oak, one-seed and alligator juniper, piñon pine,
   ponderosa pine, blue spruce, and Douglas fir.
Counties: Socorro.
Administration: Cibola National Forest—Magdalena Ranger District,

BLM—Socorro Field Office at the range's southern end, numerous
private inholdings and mining claims.

Wilderness: The 8,904-acre BLM Devils Backbone Wilderness Study Area is
centered on a sharp, jagged hogback at the range's southeastern end.

Getting there: The easiest access is via Water Canyon, which leads from
US 60 east of Magdalena to a Forest Service campground. On the
west, several Forest Roads head east into the mountains from graded
dirt NM 107 and from Magdalena. No public access exists on the
south to Sawmill Canyon. Access from the southeast is via a dirt road
starting at US 60 west of Socorro at "the box," with side roads running
to the west as far south as the Ryan Hill Canyon Trail.

You need a creative imagination to look southwest from the village of
Magdalena and see the visage of Mary Magdalene outlined in rocks and
shrubbery on Magdalena Peak's east side, especially as the 1950–52 drought
killed some of the vegetation in the configuration. Other explanations for
the name—all apocryphal—include a legend that the apparition of Mary
Magdalene saved some Mexican settlers from Indian attack by miraculously
appearing on the mountain, frightening the Indians. Yet another story
attributes the name to a Spanish priest giving a commemorative name.

But if the origin of the name is obscure, the origin of the mountains
themselves is not. The Magdalenas were formed during the general geological turmoil in central New Mexico during mid-Cenozoic time, beginning about 36 million years ago; the mountains' light-colored, fine-grained,
angular volcanic rocks are clearly akin to rocks in the San Mateo Mountains
and other mountains of the Datil-Mogollon Section of the Transition Zone.
Here, as geologist Dave Love has explained: "The central magdalenas are
part of the nest of calderas that erupted between 32 and 28 million years
ago. Basin and Range faulting has uplifted and rotated the mountain blocks.
Magdalena Peak and Elephant Mountain are more recent volcanic eruptions, approximately 13 million years ago." (personal communication).

Physiographically, the Magdalena Mountains are oriented north-south,
flanked on the east by the Rio Grande Rift and on the west by the large,
grassy basin of Milligan Gulch, which separates the Magdalenas from the
San Mateo Mountains and outlines the mountains' southern boundary.

On the north, the gap separating the Magdalena Mountains from the
Bear Mountains has been a natural transportation funnel since prehistoric

Granite Mountain, east of village of Magdalena

times. In the nineteenth and twentieth centuries, stockmen drove tens of thousands of sheep and cattle to stockyards at the village of Magdalena, then linked by rail with Socorro.

Large portions of the Magdalenas are conspicuously wild and seldom visited, yet formal wilderness protection is unlikely because of roads leading to the crest and the lightning research facilities there, as well as private inholdings that include mining claims.

The mining claims are artifacts from the prospecting fervor that followed silver strikes in the area around 1863. Far more significant was the strike in 1866 that spawned the boomtown of Kelly, three miles southeast of Magdalena. J. S. "Old Hutch" Hutchason's claim, which he sold to Andy Kelly, has been called "one of New Mexico's greatest bonanzas" (Paige W. Christiansen, *The Story of Mining in New Mexico*). By 1900 about $9 million in lead, zinc, and silver had been shipped from the Magdalena mining district. But the richest deposits soon were exhausted, and when zinc played out in the 1930s, Kelly, once home to three thousand people, began to die. Today, only sightseers visit the abandoned town.

Alligator juniper, Magdalena Mountains

Baldy Crest Trail, Magdalena Mountains

The Magdalena Mountains, like their sister ranges, are dry, though not entirely so; the only perennial streams are in Water and Sawmill Canyons, though springs are scattered throughout the mountains. Lower and south-facing slopes are sparsely vegetated with plants such as yucca, oak, mountain mahogany, and piñon. Alligator junipers (*Juniperus deppeana*) are common here, including some gargantuan specimens. Their distinctive checkered bark, compelling a comparison with an alligator's hide, is unique among trees. At higher elevations are ponderosa pines and then still higher, Douglas fir and blue spruce.

By around 1900 desert bighorn sheep had been hunted to extinction in the mountains, but as late as 1931 hikers still came across the animals' horns. Then in 1997 bighorns were reintroduced into the Magdalenas, after the BLM set up rainwater catchment tanks. Other mammals here include javelinas, pronghorns, mule deer, coyotes, tassel-eared Abert's squirrels, gray and red foxes, black bears, bobcats, and mountain lions. Birds include Merriam's turkeys, bald and golden eagles, Gambel's and scaled quail, and various owls.

In mid- to late-summer, during New Mexico's monsoon season, electrically charged thunderheads build over the Magdalenas, and some explode in lightning over South Baldy peak, famous among meteorologists for frequency of lightning strikes. At the Irving Langmuir Laboratory for Atmospheric Research, located below South Baldy's summit to the south, meteorologists and atmospheric physicists study the discharges. Near the summit are large metal lightning-attracting grids. Even on a clear day, being at the summit near the metal is an unsettling experience. In 2005, the road to the crest was paved, to allow access to the Magdalena Ridge Observatory, to be completed 2008–09, as well as to the meteorological facility, operated by the NM Institute of Mining and Technology in Socorro.

Despite this, South Baldy and its lower sibling, North Baldy, are the most popular hiking destinations, reached from trailheads at Water Canyon. The four-mile Copper Canyon Trail leads to the crest and a relatively short hike to the top of South Baldy; the hike to North Baldy along the crest is longer. The much-shorter Water Canyon Mesa Trail also affords exceptional views, especially of the Rio Grande Valley. The Water Canyon Campground is shaded and cool in summer—and at least as pleasant in other seasons, too.

Gallinas Mountains—Tres Montosas

Location: At the northeast end of the Plains of San Agustin, north of
    US 60 west of Magdalena.
Physiographic province: Datil-Mogollon Section of the Transition Zone.
Elevation range: 7,000–8,000 feet.
High point: Tres Montosas, 8,531 feet, center; 8,252 feet, north; and 8,143
    feet, east.
Other major peaks: Gallinas Peak, 8,442 feet; Lion Mountain, 8,263 feet.
Dimensions: 6.5 x 22 miles.
Ecosystems: Piñon-juniper, ponderosa pines.
Counties: Socorro.
Administration: Cibola National Forest—Magdalena Ranger District.
Getting there: From Magdalena, NM 169 heads northwest to parallel the
    range on the northeast. From NM 169, Forest Road 10 heads west
    into the mountains, eventually connecting with County Road E1

Plains of San Agustin from a Tres Montosas summit,
with the skyline of the San Mateo Mountains in the right rear

on the range's west side after crossing private land. The Tres
Montosas are on private land.

Persons driving on US 60 across the Plains of San Agustin cannot help but
notice the Tres Montosas—huge, steep-sided piles of volcanic scree rising
abruptly from the uplands just north of the highway. Drivers hardly notice
at all the Gallinas Mountains, of which the *Tres Montosas* ("three brushies")
are but an outlier. It's a large, sprawling range with only two named summits, except for the Tres Montosas: Lion Mountain and Gallinas Peak. Even
that peak's name likely is borrowed from Gallinas Spring, a watering hole
for wild turkeys, which, in New Mexico Spanish, are often called *gallinas*.
As if that weren't enough, just beyond the eastern boundary of Socorro
County is another mountain group with the same name. If mountains
could have an identity crisis, these would.

A few people hunt wild turkeys—and elk and mule deer—in these
mountains, just as did the Indians of the Mogollon Culture whose ruins are
found here. Otherwise, the Gallinas Mountains remain largely unnoticed.

No campgrounds, no trails, no peaks to climb, no streams to fish, no ghost towns or historic sites to explore, no old mining camps—just miles and miles of extremely complex, highly-dissected topography, covered almost uniformly by piñon-juniper.

Geologically and ecologically, the Gallinas are of a piece with the other low, volcanic ranges surrounding the Plains of San Agustin, the remnants of mid-Cenozoic volcanism in southwestern New Mexico that began about 36 million years ago. The mountains' closest sibling is the Datil Mountains just to the west.

Datil Mountains

Location: North of US 60, immediately northwest of the village of Datil.
Physiographic province: Datil-Mogollon Section of the Transition Zone.
Elevation range: 8,000–9,556 feet at Madre Mountain for the Datil
   Mountains.
High point: Datil Mountains—Madre Mountain, 9,556 feet; Sawtooth
   Mountains, 9,298 feet, unnamed summit at the head of Hay Canyon;
   Crosby Mountains—South Crosby Peak, 9,095 feet.
Other major peaks: Datil Mountains—Davenport Lookout, 9,354 feet;
   Sawtooth Mountains—Capital Dome, approx. 8,590 feet.
Dimensions: Datil Mountains, 12 x 25 miles; Sawtooth Mountains, 1.5 x 6.5
   miles; Crosby Mountains, 3 x 6.5 miles.
Ecosystems: Piñon-juniper, ponderosa pine, spruce, and Douglas fir.
Counties: Catron.
Administration: Cibola National Forest—Magdalena Ranger District.

The Datil Mountains are a small range at the northwest edge of the Datil-Mogollon Volcanic Field, just northwest of the Plains of San Agustin. Their kinship with other ranges in this field is revealed by the gray-pink-lavender rocks reminiscent of those in the Magdalena and San Mateo Mountains to the southeast. The Datils' closest relative is the Gallinas Mountains immediately east, across North Lake and Dog Springs Canyons and the Plains. The Cenozoic volcanoes whose debris formed these rocks erupted 31.2–28.8 million years ago. When the hot ash fell in the area of the present Datil Mountains, it was altered by heat and compression into welded, ashflow tuff, a hard rock characterized by jagged edges and holes formed by gas

Enchanted Tower, Datil Mountains

bubbles. (Unwelded ashfall tuff, a much softer rock, also is present in the area.) Eventually, the tuff and other rock layers were tilted. Erosion of these layers created the mountains and canyons we see today.

As the welded tuff cooled during this process, it often fractured vertically, and weathering along the joints has resulted in dramatic formations, including some free-standing pillars. These are especially visible in Thompson Canyon, reached by Forest Road 59, which branches north from US 60 about five miles west of Datil. Rock-climbers are especially fond of these pillars and cliffs and have given them names like Enchanted Tower, the Frog Prince, Renaissance, and the Land Beyond. The name "Datil" is Spanish, *datíl,* "soapweed yucca"; Spanish-speakers emphasize the second syllable, English-speakers the first.

Climbers also have created numerous campsites in Thompson Canyon, a charming little valley. Forest Road 59, primitive but in good weather passable by most passenger cars, climbs through the canyon to the Davenport fire lookout, from whence one can get a good idea of the range's extent and character. From here are vistas of a surprisingly large area of steep-sided hills cloaked with dense conifer forest. Even more striking is the view from the east, where a huge fault scarp has created a fortress wall of vertical cliffs. This seldom-seen view is reached by taking Forest Road 60 east over Monument Saddle. Wildlife in the Datil Mountains includes elk, mule deer, Merriam's turkeys, tassel-eared (Abert's) squirrels, black bears, bobcats, coyotes, foxes, quails, and numerous species of raptors, including golden and bald eagles.

About 1.5–2 air miles northeast of Davenport lookout is the range's highest point, Madre Mountain, an inconspicuous grassy summit. No trail leads to it, but from where Forest Road 59 makes a sharp turn from east to west at the head of Thompson Canyon it is a rather short, straightforward, cross-country hike up the ridge toward the top. Actually, no developed trails exist anywhere in the Datil Mountains, though numerous Forest Roads penetrate the hills. These roads, however, can be nightmarish for vehicles when the soil is wet.

Except for Madre Mountain, there are no named summits in the Datils.

Sawtooth Mountains

This subrange extends from the Datil Mountains west of Davenport Canyon and arches north and then west to end in several isolated formations northeast of Pie Town. Forest Road 6A branches from US 60 east of Pie Town and

Sawtooth Mountains

leads into the mountains. The Sawtooths, while smaller than the Datils, are far more scenic; the volcanic rocks here eroded into spectacular cliffs and spires that, were they located anywhere but in this remote part of the state, would be a major attraction. The only named summits in the range are the isolated formations to the west: Monument Rock, Lone Mountain, Capital Dome, and Castle Dome. Near Lone Mountain, the Continental Divide briefly enters the Sawtooth Mountains before swinging to the northwest.

While no official hiking trails exist, a good place from which to explore the Sawtooths is Forest Road 25, 4.1 miles north of US 60 on Forest Road 6A, where undeveloped campsites are located near Monument Rock and other formations. As with other mountains in the Mogollon Volcanic Complex, the Sawtooths are conspicuously devoid of surface water.

Crosby Mountains

Immediately south of US 60 from the Datil Mountains, west of the village of Datil, are the Crosby Mountains. Here a layer of basaltic andesite lava flows capped the volcanic sandstones and tuff before being eroded into the present very rugged topography. This group, while smaller in extent than the main Datil Mountains, are more rugged, with several summits fringed with vertical volcanic cliffs. The high point is South Crosby Peak, 9,095 feet in elevation. Access is via Forest Roads.

Alegres Mountain

Location: South of Pie Town, west of County Road 56.
Physiographic province: Datil-Mogollon Section of the Transition Zone.
Elevation: 10,229 feet.
Relief: 8,000–10,200 feet.
Ecosystems: Piñon-juniper, mountain mahogany, ponderosa pine.
Counties: Catron.
Administration: Private.
Getting there: Private ownership of the mountain limits public access, though County Roads 56 and 58 encircle the mountain and afford spectacular views.

Rising abruptly half a mile above nondescript hills, ringed by cliffs like battlements, isolated and solitary, and higher than any peak within fifty miles except for the Mogollon Mountains, Alegres Mountain dominates this part of New Mexico like a medieval fortress. Curious, then, that its name means "happy, bright." No one seems to know why.

Geologically, Alegres Mountain is a remnant of basaltic andesite lava flows that occurred 27.7 million years ago, accumulating in layers as much as 650 feet thick. Today Alegres Mountain appears as a very tall mesa, rising via steep cliffs to culminate in a flat, truncated summit shaped, when viewed from above, like a dog bone. In the middle of the shank is the highest point, with a secondary summit, 10,205 feet in elevation and 0.25 miles to the northeast.

About 3.3 miles to the south-southwest is Little Alegres Mountain, a near replica of its namesake—another dog-bone-shaped mesa with paired summits—but about a thousand feet lower, at 9,022 and 9,005 feet. The Continental Divide runs over both Alegres and Little Alegres; indeed, atop Alegres Mountain the Continental Divide reaches its highest point in New Mexico. But limited access to the mountain—along with a very steep, difficult climb to its top—persuaded the BLM to route the Continental Divide Trail to the west and north of the mountain itself.

Mangas Mountains

Location: North of the Plains of San Agustin, north of NM 12, 8 miles
south of the community of Mangas, south of US 60.

Physiographic province: Datil-Mogollon Section of the Transition Zone.

Elevation range: 7,200–9,600 feet.

High point: Mangas Mountain, 9,691 feet.

Other major peaks: None named.

Dimensions: 10 x 15 miles.

Ecosystems: Piñon-juniper, oak, and ponderosa pine.

Counties: Catron.

Administration: Apache National Forest (administered by the Quemado
Ranger District of the Gila National Forest).

Getting there: From NM 12, 34 miles west of Datil, well-graded Forest
Road 214 branches north to arrive after 11 miles at Valle Tio
Vences Campground, at the west base of Mangas Mountain.
The campground also can be reached from US 60 via County
Road A95, which branches south 12.6 miles west of Pie Town.
This well-graded gravel road reaches the campground after
22.4 miles, going through the old community of Mangas and up
San Antone Canyon.

A diffuse range of forested mountains and hills, the Mangas Mountains, along
with Mangas Springs and the ranching community of Mangas, owe their
name to Mangas Coloradas, "Red Sleeves," leader of the Warm Springs Apache
band in the 1800s. He certainly would have been familiar with this area, especially the springs here, an important water source in this arid region.

The mountains are a roughly circular complex, centered on the highest point, Mangas Mountain. The views from the fire tower atop Mangas
Mountain are well worth the short hike on the dirt road, normally closed to
vehicles, leading to the summit. The Mangas Mountains are among several
volcanic edifices around the Plains of San Agustin. Here one thousand one
hundred feet of basaltic andesite layers built a shield volcano. Almost all the
mountains you see from the top of Mangas Mountain have a similar volcanic origin, including the San Mateo and Luera Mountains and Pelona
Mountain to the southeast, and Horse Mountain and the Datil Mountains
to the northwest.

The Continental Divide and the Continental Divide Trail run together over the mountains. The CDT is perhaps the only hiking trail here, though Forest Roads make for pleasant walking. It's also great country for equestrians; corrals are provided at the Forest Service's Valle Tio Vences Campground. But the people most attracted to the mountains' understated but pleasant conifer forests and valley meadows are hunters, as the mountains are excellent habitat for deer, elk, wild turkey, and other game species.

Gallo Mountains

Location: In the Apache National Forest, north of NM 12, roughly bisected by NM 32, west of the Mangas Mountains.
Physiographic province: Datil-Mogollon Section of the Transition Zone.
Elevation range: 7,800–9,000 feet.
High point: Fox Mountain, 9383 feet.
Other major peaks: Gallo Peak, 9,255 feet; Agua Fria Mountain, 8,835 feet.
Dimensions: 4.5 x 17.5 miles.
Ecosystems: Piñon-juniper, oak, and ponderosa pines.
Counties: Catron.
Administration: Apache National Forest (administered by the Gila National Forest—Quemado Ranger District).
Getting there: From NM 32 in Jewett Gap, Forest Roads branch east and west into the mountain; Forest Road 770 leads to the top of Fox Mountain.

The Gallo Mountains are one small bead on the necklace of volcanic ranges around the Plains of San Agustin. They're a small, somewhat ill-defined group, trending generally east-west. Characterized by pleasant forest of ponderosa pine and Gambel oak, with a spotting of springs, the Gallos are best known to hunters. They're excellent habitat for elk and the range's namesake, wild turkeys. *Gallo* may mean "rooster" in a Spanish dictionary, but in rural New Mexico it more often means "wild turkey," along with *gallina*, "chicken."

Horse Mountain

Location: 25 air miles southwest of Datil, near New Horse Springs, northwest of NM 12.
Physiographic province: Datil-Mogollon Section of the Transition Zone.

Elevation range: 7,200–9,200 feet.

High point: Horse Peak, 9,450 feet.

Dimensions: 2 x 4 miles.

Ecosystems: Piñon-juniper, oak, and ponderosa forest.

Counties: Catron.

Administration: BLM—Socorro Field Office.

Wilderness: Horse Mountain encompasses the 5,032-acre BLM Horse
    Mountain Wilderness Study Area.

Getting there: From NM12, two county roads, B040 and B034, head
    north; these and other dirt roads branching from them provide
    access to the mountain's western and northern sections.

Part of the larger array of volcanic mountains surrounding the Plains of San
Agustin, Horse Mountain is less a single summit than a complex of canyons,
ridges, and tops. The mountain is the eroded remnant of a volcano that
erupted approximately 13 million years ago, the topography made up of
intermediate layers of dacitic lava. The whole thing rests on the eroded
remains of still older volcanoes.

Tularosa Mountains

Location: South of NM 12, at the southwest end of the Plains of San
    Agustin.

Physiographic province: Datil-Mogollon Section of the Transition Zone.

Elevation range: 7,000–9,700 feet.

High point: Eagle Peak, 9,786 feet.

Other major peaks: O Bar O Mountain, 9,417; John Kerr Peak, 8,868.

Dimensions: 15 x 27 miles.

Ecosystems: Piñon-juniper, alligator juniper, oaks, ponderosa pine,
    Douglas fir.

Counties: Catron.

Administration: Gila National Forest—Reserve Ranger District.

Getting there: Best access to the mountains is from Forest Roads branch-
    ing south from NM 12. The main Forest Roads are 28, 47, 94, and 289.

The Tularosa Mountains are among the largest of several ranges in the
Datil-Mogollon Volcanic Field. Underlying the range is a complexly-faulted,

tilted fault block. The range is approximately twenty-seven miles long, arching from Negrito Creek on the south to the Tularosa River and NM 12 on the north, where the Tularosa Mountains meet the closely-related Mangas Mountains. On the range's eastern flanks, north of Moraga Canyon, is a four-mile, east-west chain of unnamed summits labeled the Long Canyon Mountains, rising to approximately eight thousand four hundred feet.

The Tularosa Mountains take their name from the Tularosa River, which took the Southwest Spanish term meaning, "having reeds, cattails." As applied to the mountains, the name is ironic, as the range is conspicuously dry, having no perennial streams or lakes and only a few springs, making this section of the Continental Divide Trail, which stays close to the physical Divide here, especially challenging for through-hikers.

The topography is rolling rather than jagged, ponderosa forest interspersed with grassy parkland. It's excellent elk habitat, and indeed, hunting probably is the main recreational activity. The Continental Divide Trail is the sole significant hiking route, entering the southeastern mountains near Coyote Peak and hugging the Divide for approximately fifty-four miles to cross NM 12 and join Forest Road 218. The Forest Service has surveyed the route and marked it either with cairns or other markers and has constructed some tread in difficult sections, but much hiking is cross-country or over Forest Roads.

Elk Mountains

Elevation range: 7,800–9,500 feet.

High point: Elk Mountain, 9,799 feet.

Other major peaks: East Elk Mountain, 9,028 feet; Middle Elk Mountain, 8,739 feet.

Dimensions: 4 x 19 miles.

Counties: Catron.

Administration: Gila National Forest—Reserve Ranger District.

Getting there: Forest Road 30 goes through O Bar O Canyon on the range's north side, joining a county road that goes around the east of the mountains.

The Elk Mountains are geologically one with other ranges along this portion of the Continental Divide. The multiple summits in the group actually are iterations of Elk Mountain, which includes three tops of dark basaltic andesite. Not individually named, they reach, west to east, 9,678, 9,722, and 9,799 feet in elevation. These are not to be confused with other and separate Elk Mountains in the range: Elk (see above), Middle Elk, at 8,739 feet, and East Elk, at 9,028 feet. These mountains are made up of lighter-colored rhyolite, highly absorbent with little surface water.

Had you visited the Elk Mountains in the first years of the twentieth century, their name would have seemed a tragic irony, for from 1900 to 1915 elk were completely absent from New Mexico, hunted to extinction. Elk reintroduction began in 1915 and was so successful that in some parts of Catron County elk are seen as a nuisance—though not in the Elk Mountains.

O Bar O Mountains

Despite the name, this small satellite of the Tularosa Mountains is just a single mountain, O Bar O Mountain, with its 9,417-foot summit and a smaller top just to the east. The mountain is the relatively undissected remnant of a basaltic andesite shield volcano, 25–26 million years old. The name reflects a local ranch brand.

Kelly Mountains

Location: South of Reserve, northeast of Alma, southeast of the San
    Francisco River, northwest of the North Fork of Devils Creek, south
    of the Saliz Mountains.
Physiographic province: Datil-Mogollon Section of the Transition Zone.
Elevation range: 5,400–7,600 feet.
High point: Kelly Brushy, twin summits, 7,667 and 7,652 feet.
Other major peaks: None named.
Dimensions: 4 x 7.5 miles.
Ecosystems: Piñon-juniper, alligator juniper, ponderosa pine, and
    Gambel oak.
Counties: Catron.
Administration: Gila National Forest—Reserve Range District.
Getting there: From US 180 Forest Road 26 heads east, leading to Forest
    Road 32 and then Forest Road 504, which goes into Frying Pan
    Canyon and south of the mountains.

The Kelly Mountains are a small mountain group, one of many such groups in the Datil-Mogollon Volcanic Field. Like other formations in the area, they consist primarily of basaltic andesite fault blocks overlying layers of ashflow tuffs. They are aligned generally east-northeast—west-southwest and are centered on the forested twin summits of Kelly Brushy, 7,667 feet and 7,652 feet in elevation. The mountains take their name from the Kelly family of Alma, whose members still live in the area.

Saliz Mountains

Location: Southwest of Reserve, east of US 180, east of Saliz Canyon,
     west of the San Francisco River.
Physiographic province: Datil-Mogollon Section of the Transition Zone.
Elevation range: 5,400–7,500 feet.
High point: Unnamed summit, 7,581 feet.
Other major peaks: Saliz Hill, 6,751 feet.
Dimensions: 3 x 15 miles.
Ecosystems: Piñon-juniper, alligator juniper, mountain mahogany,
     ponderosa pine, oaks.
Counties: Catron.
Administration: Gila National Forest—Reserve Ranger District.
Getting there: From US 180 and NM 12 and 435, Forest Roads head
     into the Saliz Mountains.

Aligned along a forested ridge running northeast-southwest, the Saliz Mountains are among the larger of several mountain groups, often ill-defined, in this area, all originating with the geological turmoil that transformed the region about 30 million years ago. These mountains are the eroded remains of a complexly faulted horst (block separated by faults from lower blocks) of volcanics, that formed 28.1 million years ago. The mountains lie between the Reserve graben to the north and the Alma Basin to the south.

The summits along the ridge generally are forested, undistinguished, and unnamed, at least officially. The mountains are relatively gentle in the north but get steadily higher and more rugged toward the south. The name *Saliz* likely is derived from a family name, perhaps early Spanish-speaking homesteaders, but the name also could refer to *salt* (Spanish, *saliz*) or to *willows,* as similar words denote willows in Spanish and Latin. The name is ubiquitous in the area: Saliz Hill, 6,751 feet in elevation, at the range's south-west end, Saliz Pass just to the east, and narrow Saliz Canyon at the range's south end.

Brushy Mountains

Location: Immediately west of Saliz Pass, north of Alma.
Physiographic province: Datil-Mogollon Section of the Transition Zone.
Elevation range: 5,500–7,000 feet.
High point: Unnamed summit, 7,345 feet.
Other major peaks: None named.
Dimensions: 2.5 x 9.5 miles.
Ecosystems: Piñon-juniper, alligator juniper, ponderosa pine, Gambel
     and Emory oak.
Counties: Catron.
Administration: Gila National Forest—Glenwood Ranger District.
Wilderness: Almost entirely within the 29,304-acre Blue Range Wilderness,
     which abuts the much larger Blue Range Primitive Area in Arizona.
Getting there: The only vehicular access is via US 180 on the east and
     south or Forest Road 232, which runs along the north side of the
     mountains and the wilderness area.

No mountains should be named Brushy, not in a region where *all* the mountains are "brushy." In New Mexico it's like naming a creek "Dry." New Mexico has nine *Brushy Mountains*—eight in the state's southwest quad-rant—as well as features with the same name but in Spanish, *Montosa.* Just south of the Brushy Mountains, in the Sierra Aguilada (see entry), is another Brushy Mountain, 7,405 feet in elevation, while yet another Brushy Moun-tain, at 7,420 feet, is just a few miles farther south, west of Cliff. (Almost as imprecise is naming a mountain range *Blue;* almost all mountain ranges appear blue when viewed from a distance.)

Not that the Brushy Mountains aren't brushy, however. They wear a dense pelage of ponderosa pine, oak, and piñon-juniper—good brushy habitat for wild turkey, javelina, black bear, and deer. Thus, of the few recre-ationists in the Brushies, most are hunters. The infrequent hikers here choose between only two trails in the Blue Range Wilderness. The Saliz Trail goes south from Forest Road 232 to follow an old road into the wilderness, culminating after 2.5 miles atop the Brushy Mountains ridge. The 8.8-mile WS Mountain Trail follows canyon bottoms south and west from the Pueblo Park Campground toward WS Mountain and the Arizona border. Around Pueblo Park, rockhounds search for the semi-precious gemstone laboradorite, previously identified here as bytownite, a rare feldspar mineral that forms pale-yellow crystals in igneous rock, such as the Cenozoic volcanic rocks comprising the Brushy Mountains. Archaeological sites from the Mogollon Culture, a Puebloan group that lived here from AD 850 to 1350, also are nearby, hence the name Pueblo Park.

Whiterocks Mountain

In the Blue Range Wilderness, just southwest of the Brushy Mountains and higher and far more imposing, at least visually, is the massif of Whiterocks Mountain. On the north it rises south of Bear Creek and runs southwest about 6.5 miles to end at Little Blue Creek in Arizona, with most of the mountain in New Mexico. The several summits along the ridge culminate at an unnamed 8,827-foot top. From US 180 north of Alma, the "white rocks" for which the mountains likely derive their name are conspicuous.

Sierra Aguilada

Location: West of the San Francisco River and US 180, west of Glenwood.
Physiographic province: Datil-Mogollon Section of the Transition Zone.
Elevation range: 4,400–7,000 feet.
High point: Brushy Mountain, 7,405 feet.
Other major peaks: Park Mountain, 7,321 feet.
Dimensions: 5 x 13 miles.
Ecosystems: Piñon-juniper, alligator juniper, ponderosa pine.
Counties: Catron.
Administration: Gila National Forest—Glenwood Ranger District.
Getting there: From Alma, Forest Road 106 heads west then southwest to arrive after ten miles at a junction, from which another Forest Road runs east, leading to the electronic towers atop Brushy Mountain.

"Eagle Range" is the meaning of the name *Sierra Aguilada,* and the reason cited by the proposal to the U.S. Board on Geographic Names for making the name official is that eagles are indeed common among the mountains.

The Sierra Aguilada is characterized by steep, forested slopes rising toward rounded tops, with some cliffs near the river. The rocks are 24 million-year-old andesite on top of 26 million-year-old dacite. It's an outlier of the larger amorphous complex of mountains west of the San Francisco River and US 180. All share a common origin in mid-Cenozoic volcanism beginning around 36 million years ago.

Mule Mountains

Location: West of US 180, south of NM 78, east of the village of Mule Creek.
Physiographic province: Datil-Mogollon Section of the Transition Zone.
Elevation range: 5,600–6,000 feet.
High point: Unnamed summit, 6,321 feet.
Dimensions: 2.5 x 4 miles.
Ecosystems: Piñon-juniper, oak, ponderosa pine.
Counties: Grant.
Administration: Private.
Getting there: From US 180, follow NM 78 for 2.5 miles to where a dirt county road branches south. After 2.25 miles, another dirt road leads southwest then north into the mountains. Respect all signs and private property.

This small uplift is an outlier of the larger, better-known complex of ranges in the Gila Wilderness. Like the others the Mule Mountains were part of the mid-Cenozoic volcanism in the region beginning about 36 million years ago. The Mule Mountains are one of several rhyolite domes in the area, dating to 13 million years ago. Other domes to the north are known for their obsidian pebbles referred to as Apache tears. The small group is characterized by gentle slopes on low hills, with few trees.

San Francisco Mountains

Location: West of Reserve, north and west of NM 12 and US 180, bounded on the north by the San Francisco River, extending into Arizona as far as Blue Creek.
Physiographic province: Datil-Mogollon Section of the Transition Zone.
Elevation range: 6,200–8,200 feet.

High point: Unnamed summit, 8,441 feet.

Other major peaks: Monument Mountain, 8,401 feet; Black Bull Peak, 8,361 feet; South Mountain, 8,343 feet; Prairie Point Peak, 8,335 feet; Leggett Peak, 7,939 feet.

Dimensions: 9 x 21 miles.

Ecosystems: Piñon-juniper, oak, ponderosa pine.

Counties: Catron.

Administration: Apache National Forest (administered by the Gila National Forest—Gila Ranger District).

Getting there: US 180 bisects the range south of Luna, with Forest Road 209 heading south and Forest Road 39 heading north. Forest Roads also go around the range's north end, along the San Francisco River.

The San Francisco Mountains are a long, linear plateau, another of the region's tilted fault-block ranges. The northeast-southwest trending plateau is hardly level but certainly more so than the steep, rugged, highly-eroded canyons and ridges that rise to meet it. Cenozoic volcanism produced the topography here, as well as the feature for which the mountains are best known by the public—San Francisco Hot Springs, on the San Francisco River at the range's north end.

Mogollon Mountains

Location: Along the western and southwestern rim of the Gila Wilderness, southwest of the Plains of San Agustin.

Physiographic province: Datil-Mogollon Section of the Transition Zone.

Elevation range: 4,700–10,000 feet.

High point: Whitewater Baldy, 10,895 feet.

Other major peaks: Willow Mountain, 10,783 feet; Mogollon Baldy Peak, 10,770 feet; and Center Baldy, 10,565 feet.

Dimensions: 40 x 60 miles.

Ecosystems: Piñon-juniper, ponderosa pine, Douglas fir, aspen, Engelmann spruce, and subalpine fir.

Counties: Catron, Grant.

Administration: Gila National Forest—Glenwood Ranger District.

Wilderness: Gila Wilderness, 557,873 acres.

Getting there: NM 159 heads east from Glenwood into the mountains, to connect with several Forest Roads; NM 435 heads south and then east from Reserve, again to connect with Forest Roads.

The Mogollon Mountains are the western rampart of the vast Gila Wilderness, a great wall of peaks and ridges arching sixty miles to separate the wilderness's interior from the river valleys to the west. By any standards, these are impressive mountains. When viewed from the west, especially from US 180 south of Glenwood, the mountains burst abruptly from the floodplain to rise a vertical mile, tier upon tier of ridges and vertical volcanic columns, to summits tall enough to capture snow even in this southerly part of the state. Numerous tops here exceed 10,000 feet, culminating in 10,895-foot Whitewater Baldy. The nearest taller mountain is 11,301-foot Mount Taylor—150 miles away.

Impressive also is the wilderness the Mogollon Mountains help define. The Gila Wilderness not only is New Mexico's first and largest wilderness, twice as large as the Pecos Wilderness, but it also was the world's first area anywhere to be set aside solely to preserve its wildness. When the Gila Wilderness was established in 1924, it encompassed seven hundred seventy-five thousand acres, but in 1933 the wilderness was cleaved by the North Star Road east of the Mimbres River, and the area east of the road became the appropriately named Aldo Leopold Wilderness (see *Black Range*).

Physiographically the Mogollon Mountains are transitional between the Colorado Plateau and the Basin and Range Provinces. They're bounded on the west by the Mangas graben and the graben of the San Francisco River valley, and on the northeast by the Plains of San Agustin graben, though the elevational contrast on the north is much more modest.

The Mogollon Mountains, along with a section of the Black Range, include almost all the Gila River's vast drainage area. By the time the Gila River leaves the mountains, it has enough water to continue flowing all the way to its rendezvous with the Colorado River in southeastern Arizona. (Contrast that to the nearby Mimbres River, whose waters rarely make it even to Deming.)

In addition to uplift between two areas of subsidence (the grabens), the Mogollon Mountains are part of the volcanic highlands between the Bursum caldera, formed 28 million years ago, and the Twin Sisters caldera, 31.4 million years ago. The Mogollon Mountains are the southwestern rim of the Bursum caldera. New Mexico has seen scores of volcanoes during its geologic

Spire, Gila Wilderness

history, many of epic scale, but few, if any, equaled the eruptions of south-western New Mexico. One observer said, "Mount Saint Helens would be a mere firecracker in comparison to the vast outpouring that took place here." Layer upon layer of ash and lava, blanketing an enormous area. Even now, millions of years later, the volcano's caldera is visible on satellite images.

The high summits here capture significant moisture, both from winter snow and summer monsoon rains. The forest here is old-growth forest, mostly of subalpine fir and Engelmann spruce. Other species include aspen, corkbark fir, Mexican white pine, and Douglas fir. From the high peaks the forest grades downward to ponderosa forest, then piñon-juniper, and eventually the grasslands of the Plains of San Agustin and the San Francisco valley.

Wild country. New Mexico's last grizzly was killed on Rain Creek here in 1931. The mountains were a redoubt for Mexican wolves. In the market-hunting orgy of the late nineteenth and early twentieth centuries that decimated wildlife statewide, the Mogollon Mountains lost not only the grizzly and Mexican wolf, but also Merriam's elk and the Gila turkey.

But the Gila trout have survived, albeit barely. The Gila trout is one of three species of trout native to either New Mexico or Arizona, and the only one common to both states. (The others are the Rio Grande cutthroat and the Apache trout.) From 1900 to the 1960s, over-fishing, introduction of non-native trout, stream degradation from logging and grazing, and other factors caused the Gila trout to go from inhabiting the entire upper Gila watershed to barely surviving in just five, small, isolated headwater streams in the Gila and Aldo Leopold Wildernesses—a ninety-five percent reduction. By 1994, reintroduction and management had increased the streams to thirteen. Encouraging, but . . .

Human natives have fared even worse. The earliest distinct prehistoric people were those of the Mogollon Culture, AD 850–1350. These people had connections to the Ancestral Puebloans to the north, and like them they left reminders of their presence in small masonry structures and dwellings, best seen at Gila Cliff Dwellings National Monument. But the Mogollon people left few clues as to why they left or what became of them. No existing tribe traces its ancestry to them.

They were gone by the time Apaches arrived. Always a mountain people, the Apaches took readily to the Mogollon Mountains, which offered food, water, and refuge—the three necessities of Apache life. As American settlers, prospectors, ranchers, and the U.S. Army began penetrating the mountains, conflict was inevitable. In December, 1885 at Soldier Hill south of Glenwood, Geronimo ambushed U.S. Army troops. Then he and his warriors made a strategic retreat up a ridge of Mount Baldy in the Mogollon Mountains. Still being pursued, he used his knowledge of the terrain to elude the troops and enter a secret stronghold in what is now called Teepee

Canyon, where he spent the winter of 1885–86. When Jack Stockbridge visited the site in 1900, remains of the camp were still visible. The engagement was one of the last battles of the Apache wars, and that winter in the Mogollon Mountains was the last Geronimo spent as a free man.

But even with the Apaches gone, the Gila Wilderness saw few travelers, and no settlements.

That was not for want of trying, however; prospectors ranged throughout the mountains, spinning off the usual tales of rich veins found and then lost. Captain Mike Cooney, a Civil War veteran, spent the latter part of his life in lonely quest for a rich vein; in 1914 his body, propped against a tree, was found in the canyon later named for him, his searching ended.

Yet Cooney did have reason to believe that someone could stumble upon rich metal; after all, his brother had done it. In 1870 Sergeant James C. Cooney was a scout for a patrol of the Eighth U.S. Cavalry looking for Apaches along Mineral Creek when he chanced upon rich silver ore. Soon after his discharge from the Army in 1876, Cooney and two associates returned to the site, relocated the ore, staked a claim, and began working it. The Apaches, however, had not yet abandoned the mountains, and in 1880 they killed Cooney. His friends, including his brother, used ore from the mine to build a tomb, still a local landmark. The mine survived its discoverer and around it grew a vigorous mining camp, which adopted Cooney's surname. At one time Cooney had six hundred residents, three stamp mills, a school, a church, two hotels, and a post office. But declining metal prices and a flood in 1911 doomed the town; only foundations now remain.

The nearby mining camp of Mogollon, however, has clung to life. In 1889, John Eberle got lucky with his Last Chance Mine. Eberle named the resulting boom town for the mountains. (The mountains had much earlier been named for Juan Ignacio de Flores Mogollón, New Mexico governor from 1712 to 1715, who led an expedition here and very likely also did some prospecting.)

Today, recreation is the main activity in the Mogollon Mountains; the Gila Wilderness has six hundred miles of hiking trails, and the high peaks are an obvious destination. The Crest Trail, No. 182, leads to Hummingbird Saddle (all visitors concur that hummingbirds do frequent the site), Whitewater Baldy, and Mogollon Baldy. The trailhead is reached by NM 78, which runs east from US 180 three miles north of Glenwood, passing through the old mining town of Mogollon to reach the Sandy Point trailhead.

Jerky Mountains

Location: In the Gila Wilderness, east of the West Fork of the Gila River,
  south of Iron Creek.

Physiographic province: Datil-Mogollon Section of the Transition Zone.

Elevation range: 6,800–9,000 feet.

High point: Unnamed top, 9,033 feet.

Other major peaks: Lilley Mountain, 8,949 feet.

Dimensions: 5 x 8 miles.

Ecosystems: Piñon-juniper, alligator juniper, ponderosa pine, Douglas fir.

Counties: Catron.

Administration: Gila National Forest—Glenwood Ranger District.

Wilderness: Gila Wilderness, 557,873 acres.

Getting there: From Forest Service campgrounds along Willow Creek in
  the north, several trails head south into the mountains. Around
  Turkeyfeather Pass are major trail junctions, with Trail 164 leaving
  the valley to reach the crest of the Jerky Mountains.

In the early 1880s, John C. Lilley joined a handful of other settlers in attempt-ing to homestead in the vast Gila country. At that time it was a dangerous place; Geronimo had not yet surrendered, and Apaches, in the last days of their resistance, still roamed the Gila Country. Lilley settled at one of the more remote sites in the wilderness, the little park that bears his name near Lilley Park Spring in the shadow of Lilley Mountain, the only named summit in the Jerky Mountains. And it was in the wilderness on Clear Creek in December 1885, one year before Geronimo's surrender, that Lilley and his neighbor, Thomas C. Prior, were killed by Apaches. They were buried there. In the same raid, another neighbor, Presley "Papenoe" Papence, also was killed. They added their names to the list of other wilderness homesteaders killed by Apaches: the two McKenzie brothers, William Benton, a man named Baxter, James C. Cooney, and many others.

Yet many of the violent deaths among early settlers resulted not only from Apaches but also from feuds and grudges among the settlers them-selves. These included the Grudging brothers, Bill and Tom, with whom orig-inated the name Jerky Mountains. They had a cabin near a spring on the mountains' south side that they used as a base for market hunting. At the cabin they would dry the meat, or "jerk" it, then pack out the dehydrated meat to sell. They became embroiled in a dispute with another settler, Tom

Wood. They ambushed and killed Wood's son and a Mexican companion. Wood then sought revenge upon the Grudging brothers, killing Bill in the Gila Wilderness and following Tom to Louisiana and killing him there.

It's ironic that the Gila Wilderness of today, a place of beauty, peace, and tranquility, should have such a violent past.

The geology of the Jerky Mountains is similar to that of the surrounding area, consisting mainly of rhyolitic rocks that have filled the Bursum caldera, a volcanic feature dating from 28 million years ago.

Diablo Range

Location: Southeast of the Mogollon Mountains in the Gila Wilderness, west of the Gila River, north of Turkey Creek.
Physiographic province: Datil-Mogollon Section of the Transition Zone.
Elevation range: 6,000–8,400 feet.
High point: Unnamed summit, 8,820 feet.
Other major peaks: Granite Peak, 8,731 feet, White Pinnacle, 8,730 feet.
Dimensions: 5 x 13 miles.
Ecosystems: Ponderosa pine, Douglas fir, white fir, and blue spruce.
Counties: Catron.
Administration: Gila National Forest—Wilderness Ranger District.
Wilderness: The range is entirely within the 557,873-acre Gila Wilderness.
Getting there: From the Gila Visitor Center, near the north end of NM 15, Trail 160 heads west into the range. Once there it joins Trails 161, 162, and 155 within the mountains.

The Diablo Range is among several ill-defined mountain groups folded into the vast topographic complexity of the Gila Wilderness. In this range's northeastern foothills, near McKenna Spring, is the point in New Mexico farthest from a road or residence (approximately ten miles). Trails in the Diablo Range generally are along stream drainages and typically lead to other places in the wilderness and not to the range itself, though Trail 150 branches from Trail 155 to lead to the site of the former lookout atop Granite Peak.

The peak's name reflects a mistake by early settlers, who seeing light-colored, welded rhyolite tuff—a common volcanic rock—at the mountain's base, mistook it for granite, another molten igneous rock but one that cools much slower than rhyolite and forms larger crystals.

Black Range

Location: Runs north-south from the Plains of San Agustin, southwest of the San Mateo Mountains and east of the Luera Mountains, ending in the south at the Cookes Range, a subrange; parallels the Rio Grande Valley about 25 miles west of Truth or Consequences.

Physiographic province: Datil-Mogollon Section of the Transition Zone.

Elevation range: 6,000–10,000 feet.

High point: McKnight Mountain, 10,165 feet.

Other major peaks: Reeds Peak, 10,015 feet; Hillsboro Peak, 10,011 feet; Diamond Peak, 9,850 feet; Sawyers Peak, 9,668 feet; Lookout Mountain, 8,872 feet.

Dimensions: 30 x 100 miles.

Ecosystems: Chihuahuan Desert shrubs and grasses, mountain mahogany, piñon-juniper, ponderosa pine, rising to spruce-fir and Douglas fir forest and extensive stands of aspen.

Counties: Grant-Sierra, Catron.

Administration: Gila National Forest—Black Range and Mimbres Ranger Districts.

Wilderness: The 202,016-acre Aldo Leopold Wilderness, New Mexico's third largest, straddles the crest at the range's north end.

Getting there: The Black Range is most easily accessed from NM 152, connecting the Rio Grande Valley with the Mimbres Valley, passing through Hillsboro and Kingston and traversing the mountains at 8,228-foot Emory Pass. Several Forest Service campgrounds are along this route. NM 59 passes through the range's northern end, connecting the Rio Grande Valley with the Forest Service's Beaverhead Work Station; the pavement ends there, though several long dirt Forest Roads continue west, northwest, and south. On the west, NM 61 and NM 35 roughly parallel the range's southern extent along the Mimbres River; and several dirt Forest Roads head eastward toward the mountains. From the east, most access to the mountains is blocked by private land, except for roads around Winston and Choride.

The Black Range and neighboring Gila Country encompass the largest and most remote area of wild land in New Mexico. The range's foothills and approaches are all but uninhabited; Forest Roads are few and usually rough, and even "major" trails are faint. One of New Mexico's last grizzlies was killed in the Black Range in the 1930s, but other animals, extirpated else-where, have survived here, sometimes very well. (In 1990 one of the rare, recent sightings of a jaguar in New Mexico occurred in the Black Range.) If the Apache leader Victorio somehow were to return to his stronghold in the Black Range, he'd notice a few changes—but not many—and in most of the range, none at all. He might wonder, "Where have all the prospectors gone, all the soldiers? Where have my people gone? Why are the mountains so empty, so silent?"

A slender, linear range, the Black Range is second only to the Sangre de Cristo Mountains in length. On the east, canyon-cut plains sloping upward from the Rio Grande grade into mesas, then into rugged foothills, and finally into steep canyons and ridges culminating at the crest. From the west the range ascends somewhat more decisively from the Mimbres River Valley. On the north, the range becomes attenuated as it approaches the Plains of San Agustin, while 7,932-foot Thompson Cone marks the south-ern boundary. Older maps often label the subrange south of Emory Pass as the Mimbres Mountains, the name likely inspired by the Mimbres River, whose Spanish name simply refers to willows along it.

Geologically, the Black Range is the faulted center of the Emory caldera, a 34.9 million-year-old resurgent caldera and pluton later cut by Basin and Range faults. As such, it is part of the Datil-Mogollon Section of the Transition Zone, though its core of Precambrian igneous and Paleozoic sed-imentary rocks probably controls the steep topography as much as the overlying mid-Cenozoic volcanic deposits. The Black Range, the San Mateos, the Magdalenas, the Datils, and numerous other southwestern mountain groups all have a common origin—and a similar character to their rocks.

A characteristic the Black Range shares with its neighbors in the Datil-Mogollon Section of the Transition Zone is that its rocks and soils readily absorb moisture. Springs occur occasionally, but surface water is rare. The most significant streams are the Mimbres River, which heads near Reeds Peak, and Las Animas Creek and Percha Creek. Most other streams are intermittent.

The mountains' vegetation reflects both this aridity and the southern latitude. Even at elevations above nine thousand feet Chihuahuan Desert plants such as yucca, cacti, and mountain mahogany are found on south-facing slopes. Elsewhere are large, dense stands of Gambel oak, as well as

The Black Range, from the crest

piñon-juniper forest. In cooler, wetter microclimates are ponderosa pines and western white pines, while still higher are spruce and fir. Around Iron Springs are maples that turn red in the fall.

Early Hispanics called the Black Range *Sierra Diablo,* "devil mountains." Several explanations are plausible, but it's tempting to say the name reflects the range being "devilishly" rough. The Black Range name resulted from the mountains having a distinctly dark appearance—black under some lighting conditions—from the mountains having a dense coat of dark-hued coniferous trees, especially spruce and Douglas fir.

The steepness of slopes here results from the range's origin. During the mid-Cenozoic upheaval throughout southwestern New Mexico, beginning about 36 million years ago, the range's Precambrian granite core and overlying Paleozoic sediments were intruded by igneous rock, so that today the range consists of uplifted layers dropping steeply from both sides of the volcanic crest.

As early as 300 BC people of the Basketmaker Culture lived along the Mimbres River and in the western foothills of the Black Range. They crafted decorative tin beads from cassiterite pebbles found in streambeds in the mountains. By AD 850 the Basketmakers had evolved into the Puebloan culture named Mimbreño, part of the larger Mogollon group. The Mimbreño and Jornada Mogollon dominated southern New Mexico just as the Ancestral Puebloans of Chaco Canyon did in the north. The Mimbreños settled in a dense array of small villages on alluvial benches above the river bottom. At higher elevations, their sites are rare.

Then around AD 1350 they abandoned the region that for at least a thousand years had been their home. Why they left and what became of them is still debated, but while they were here, they created pottery whose bold and beautiful painted designs still resonate, even in the sensibilities of an alien people.

About a hundred years after the Mimbreños left, the Apaches arrived, followed in 1540 by Europeans. The Apaches weren't interested in the mountains' minerals—specifically gold and silver—but they soon learned the Europeans were obsessed with them. Hispanic villagers along the Rio Grande had avoided the mountains, but there was no stopping the English-speaking prospectors once they got a whiff of a strike. In 1877 gold was discovered in the Animas Hills, a small but rugged group surrounding Copper Flat near Hillsboro; two prospectors picked up promising float at what became the Opportunity Mine. Placer gold was discovered the same year. During the winter of 1877–78, a miner named George Wells is said to have recovered ninety thousand dollars in gold dust and nuggets in Wicks Gulch. From 1877 to 1931, the Hillsboro Mining District yielded metals worth $6.9 million.

In 1878, silver was discovered in the Black Range foothills near Lake Valley. The deposit known as the Bridal Chamber was among the world's richest silver concentrations. It yielded 2.5 million ounces of silver from ore sometimes so pure it was cut with saws in the mine. But on the day the Bridal Chamber was discovered, George Daly, the mine's owner, was killed by Apaches. James McKenna's book of mining era reminiscences, *Black Range Tales,* is filled with accounts of Apache attacks and ambushes.

Consequently, the mountains saw numerous clashes between the Apaches and the U.S. Army. One such occurred September 18, 1879, near the mouth of Animas Canyon, when Victorio's band ambushed troops of the Ninth Cavalry. The graves of soldiers who died in the battle still line up in military formation in a nearby meadow. Other graves nearby are less easily explained.

Aldo Leopold

How appropriate that New Mexico's wildest wilderness area bears the name of the man who pioneered the concept of preserving wilderness and who worked and traveled here. Leopold is regarded as the father of wildlife ecology, but his influence was much more far-reaching than just that discipline. It was Leopold who articulated for much of the world the concept of a land ethic, and it was Aldo Leopold who spearheaded the drive to create the world's first formally-protected wilderness area, the Gila Wilderness of New Mexico, which at the time included the Black Range and the present Aldo Leopold Wilderness.

Aldo Leopold, born in Iowa, joined the U.S. Forest Service in 1909 and was assigned to the Apache National Forest, which then straddled the New Mexico-Arizona border and included much of the Gila Country. Leopold moved up rapidly in the Forest Service; in 1911 he became deputy supervisor of the Carson National Forest, then supervisor in 1912, the same year he married Estella Bergere from a pioneer New Mexico family. Among his employees on the Carson was a young man, about the same age, who was to have a similar evolution as a wildlife conservationist and wilderness advocate: Elliott Barker. The two became lifelong friends.

Leopold came in contact with others who shared his growing concern about wilderness preservation, and with his friend, Fred Winn, supervisor of the Gila National Forest, he mapped the boundaries of the world's first area to be set aside solely to preserve its wild character.

In 1924, the same year the Gila Wilderness was created, Leopold left the Southwest to join the U.S. Forest Products Laboratory in Madison, Wisconsin. In 1928 he began teaching at the University of Wisconsin. There he wrote his seminal *Game Management* (1933), and soon thereafter he became first chair of the newly created Department of Game Management.

The mining frenzy spawned several camps; most, such as Grafton, Robinson, Lake Valley, and Phillipsburg, were short-lived. Today, only Hillsboro, Kingston, Winston, and Chloride survive, far smaller and quieter than one hundred years ago. Like gold flakes in a prospector's pan, occasional tourists yield a few dollars for the local economies.

But even tourism is modest here. The Black Range certainly has not

*Aldo Leopold knew
these mountains*

Then in 1935 he and his family purchased a worn-out farm in what was known as Wisconsin's "sand counties." Here Leopold put into practice many of his land steward-ship ideas. And here he crafted many of the essays that articulated his land ethic philosophy. The most famous collection of essays he titled *A Sand County Almanac.* He prefaced the work by saying: "There are some who can live without wild things and some who cannot. These essays are the delights and dilemmas of one who can-not." In it he wrote: "That land is a community is a basic concept of ecology, but that land is to be loved and respected is an extension of ethics."

The Aldo Leopold Wilderness isn't New Mexico's largest—it's half the size of the adjacent Gila Wilderness and slightly smaller than the Pecos Wilderness—but while the Aldo Leopold Wilderness encompasses only a part of the Black Range, few people would argue that the rest of the range isn't comparably wild. Indeed, far fewer people are to be found in the Black Range today than were here a hundred or more years ago. In the range's vastness are still to be found all the unspoiled natural beauty and solitude that inspired Aldo Leopold almost a century ago.

developed into a hiking mecca. Whether approached from the east or the west, the mountains wear a dark, forbidding cloak of dense forests on relent-lessly steep slopes. Nor is the impression dispelled within the range; on both sides of the Black Range crest from Emory Pass to Hillsboro Peak, ridges, canyons, and cliffs plunge dauntingly downward.

Trails do exist here, however, and people do hike them. The most popular trail is the ten-mile round-trip along the Black Range Crest from Emory Pass to 10,011-foot Hillsboro Peak. The gradients are relatively easy, and the vistas are spectacular, especially from the fire tower there. Were you to continue north on Trail 79 from Hillsboro Peak, you would reach the McKnight Road after 8.5 miles. From there, it's another 12.3 miles along the crest via Trail 79 to Reeds Peak and the junction with the Continental Divide Trail, which follows the Divide north along the crest. Slowly but steadily the Forest Service, working with the Continental Divide Trail Alliance and volunteer groups such as AmeriCorps and NM Volunteers for the Outdoors, is constructing tread for the Continental Divide Trail, but funds are scarce, and the logistics formidable.

Experienced long-distance hikers on the CDT often report that they weren't really prepared for just how difficult is the twenty-five-mile stretch from Reeds Peak to Lookout Mountain, far more so than a glance at the map would suggest. This is because of the impact of forest fires on the vegetation. Taking packs off, crawling under charred tree trunks blocking the trail, putting the packs back on, looking for the faint trail, encountering more burned trees and dead snags—tough, slow going. Also, the high points of the Black Range act as lightning rods, and year after year the forests are ablaze. Dense stands of successional species, such as short-lived aspens, also impede hiking.

But the same fires that have made much of the Black Range crest a thicket of dead and dying trees and a dense tangle of new growth, also have made the Black Range a laboratory for fire succession in southwestern mountains. The McKnight Burn is the best known. It occurred in 1951 and was ignited by human carelessness. It burned thirty thousand acres. The fire occurred during an especially dry spring; natural fires, sparked by lightning, typically occur during the late-summer monsoon season and thus are accompanied by rains that dampen the fire danger. The unusually intense McKnight fire destroyed protective ground cover, exposing bare soil. Now vast stands of aspen have appeared, part of the natural succession; the climax spruce-fir forest will not return for fifty or more years. In the meantime, the aspens now maturing will make the Black Range among the Southwest's premier fall foliage spectacles, though as of this writing, no hotel chains have plans to build in Kingston.

It's ironic that in these mountains, where most destinations are difficult to reach, the range's highest summit is relatively easy. From NM 35 north of

Mimbres, take the McKnight Road, Forest Road 152, as it heads generally northeast. It climbs steadily and gets progressively rougher, but barring a wet surface, it's usually passable for most vehicles with reasonably high clearance. The goal is Signboard Saddle and the junction with the Crest Trail, No. 79. The road and the trail continue north together for another mile until they reach a cleared parking area. The road veers downhill left for 0.25 mile to end at McKnight Cabin Spring, the area's only water source. From the clearing, Trail 79 leads north to the top of McKnight Mountain and beyond.

Vast and wild, the Black Range is conspicuously rich in wildlife. The species list is familiar, but in few other areas does one have such a strong sense of wildlife abundance. Wild turkeys especially seem more common here than elsewhere. Once, while camped on Animas Creek, I actually yelled at wild turkeys to keep quiet because their morning gobbling kept me from sleeping. That's the Black Range.

Pelona Mountain

Location: South of the Plains of San Agustin, north of NM 163.

Physiographic province: Datil-Mogollon Section of the Transition Zone.

Elevation: 9,212 feet.

Relief: 1,600 feet.

Ecosystems: Piñon-juniper, ponderosa pine, and grama and bluestem grasslands.

Counties: Catron.

Administration: BLM—Socorro Field Office.

Wilderness: The mountain is within the 68,761-acre Continental Divide Wilderness Study Area. Within the WSA, the area immediately around Pelona Mountain has been designated an Area of Critical Environmental Concern (ACEC), from which cattle are excluded.

Getting there: The only legal access is by foot, likely over the Continental Divide Trail. From the east this is reached from NM 163, an all-weather dirt road that branches south from US 60. From the west access is via Forest Roads 28 and 551, south of NM 12.

The main appeal of Pelona Mountain is its isolation. The remnant of a basaltic andesite shield volcano that erupted 26.6 million years ago, Pelona sits as on a dais amid vast, grassy uplands. A few unauthorized roads lead to the mountain's base, but otherwise the peak is accessible only by hiking. The Continental Divide and Continental Divide Trail pass over the mountain as they follow an east-west ridge. It's a long hike, with jagged volcanic fragments always underfoot, but the scenery is spectacular—especially the view from the summit. From there at night, the center of a circle more than one hundred miles in diameter, not a single light is visible. One's first sighting of elk here seems only... natural. But if wildlife is common here, water is not. Springs are rare, perennial streams non-existent.

The name *Pelona* (Spanish, "bald") is a relic of the days when fire or logging had removed trees from the summit. To the envy of bald men everywhere, this "baldy" has regrown its pelage.

Luera Mountains

Location: At the southeast end of the Plains of San Agustin, just north of
   the Continental Divide across Alamo Canyon.
Physiographic province: Datil-Mogollon Section of the Transition Zone.
Elevation range: 7,000–8,800 feet.
High point: Luera Peak, 9,482 feet.
Other major peaks: None named.
Dimensions: 10 x 13 miles.
Ecosystems: Piñon-juniper, including alligator juniper.
Counties: Catron.
Administration: State.
Getting there: From NM 163, several dirt roads branch northwest into the
   Luera Mountains. Respect all signs, gates, and private property.

The Luera Mountains are a roughly circular range, whose complex system of ridges and canyons rises to the central hub of Luera Peak. About 28 million years ago, a basaltic andesite volcano erupted here; in the time since then, the volcano has lost much of its original shape through faulting.

Like their neighboring mountains, the Luera Mountains are remote, little known, seldom visited, and generally lacking surface water. Piñon-juniper and ponderosa pine forests and high, open grasslands are the dominant vegetation. Except for Luera Spring, 1.8 miles northeast of Luera Peak, no other springs appear on the map, though ranchers have installed wells and water tanks. The name Luera refers to a local ranching family.

As elsewhere, hunting is the main recreational use of the mountains. No trails exist, though the Continental Divide Trail passes just south of the Luera Mountains. The top of Luera Peak is reached by a dirt road leading to electronic towers.

San Mateo Mountains

Location: Southwest of the village of Magdalena and the Magdalena Mountains, east of the Black Range.

Physiographic province: Datil-Mogollon Section of the Transition Zone.

Elevation range: 7,000–10,000 feet.

High point: West Blue Mountain, 10,336 feet.

Dimensions: 20 x 40 miles.

Other major peaks: Blue Mountain, 10,309 feet; Vicks Peak, 10,252 feet; San Mateo Mountain, 10,145 feet; San Mateo Peak, 10,139 feet; Mount Withington, 10,116 feet; Apache Kid Peak, 10,048 feet.

Ecosystems: Semi-desert grassland rising in elevation through piñon-juniper, ponderosa pine, Douglas fir, and finally aspen and spruce.

Counties: Socorro.

Administration: Mostly Cibola National Forest—Magdalena Ranger District.

Wilderness: Withington Wilderness, 18,870 acres, in the north; Apache Kid Wilderness, 44,650 acres, in the south.

Getting there: From the north, Forest Road 549 leads from US 60 to the Hughes Mill and Bear Trap Campgrounds and connects with Forest Road 138, going to the top of Mount Withington. From the west, NM 52 runs south from US 60, and from this, several Forest Roads head east into the mountains. From the east, Forest Road 225 goes west from NM 107 and leads to Springtime Campground. From the south Forest Road 139 connects with Forest Road 285 leading to the Luna Park Campground and Forest Road 225.

I remember approaching the San Mateo Mountains once from the northeast. As I glimpsed pronghorn on the vast grasslands, I had the sense of seeing the landscape as a Paleo-Indian would have seen it, ten thousand years ago. Easily I could imagine giant elk, giant bison, dire wolves, saber-toothed lions, perhaps even a Columbian mammoth—grazing in the grassland basins surrounding the mountains. In the cooler, wetter Pleistocene, such a Paleo-Indian would have stood atop what is now Mount Withington and beheld the basin now known as the Plains of San Agustin filled with water, the brisk, westerly wind pushing waves onto shores of the San Mateo foothills. In the south, Alamosa Creek would have flowed without interruption to the Rio Grande. And the tiny, threadlike waterfall in Garcia Canyon would have thundered through the narrow slot, had anyone been there to hear.

The San Mateos evoke a sense of remoteness in space as well as in time—and for good reason. No community is near the San Mateo Mountains—unless you count Monticello-Placitas, old but still viable farming communities on the mountains' southwest side, or Dusty, on the northwest; in 2001 it had one residence.

The San Mateos always have been remote. Prospecting and mining here were modest at best. The main mining camp in the San Mateos was the tiny settlement of Rosedale. W. "Jack" Richardson and his wife, Rose, discovered gold on a creek on the mountains' eastern flanks. Their mine, and the little camp that germinated nearby, took Mrs. Richardson's name. But the gold and silver veins were stingy. The camp never had more than a few cabins, though it had a post office, and when mining finally ceased in the 1930s, Rosedale died soon thereafter, leaving only rusting metal and decaying timbers. Other ephemeral camps included San Jose east of Vicks Peak and Taylor (or Ojo Caliente No. 2) on the west.

Earlier, Apaches had discouraged settlement in the San Mateos. The mountains were within the territory of the Warm Springs Apache band, Geronimo's band, named for their association with the thermal springs at Ojo Caliente near the head of Alamosa Creek. It was there that the U.S. Army in 1874 established the Warm Springs Apache Reservation. In 1875–76 as many as two thousand Apaches were at the reservation, but by 1877 most of them had decamped. That same year Geronimo, fleeing the San Carlos Reservation, was recaptured there. Two years later, Victorio's band attacked the garrison. But in 1881 Victorio was dead (Vicks Peak recalls his presence), and the Apache Wars were ending. In 1882 Ojo Caliente was abandoned. Yet one final Apache drama remained to be played out in the San Mateos.

In 1889, a White Mountain Apache who was an Army scout in the campaigns against Geronimo escaped en route to prison, having been convicted of wounding an American Army scout who'd earlier befriended him. With Apaches and Americans both against him, the young Apache turned renegade.

San Mateo Mountain (left) and Vicks Peak (right)

He was called the Apache Kid. For four years he lived the old Apache life of raiding and rustling. Finally, in September, 1894, New Mexico cattleman Charles Anderson and some cowboys ambushed rustlers in the San Mateo Mountains, killing one—the Apache Kid. He was buried near Apache Kid Peak, 10,408 feet in elevation, located in the heart of the wilderness between Blue Mountain and San Mateo Peak. The Apache Kid Wilderness, administered by the Cibola National Forest, was designated in 1980 and includes 44,650 acres.

Indians lingered in the San Mateos, however. Aldo Leopold, writing about the uncharacteristically poor results of a 1919 hunting trip to the San Mateos, blamed "meat-hunting" Indians for the paucity of game. Leopold didn't identify the tribe, but most likely they were Apaches, and in a sense, the Chiricahua Apaches and the San Mateos are still joined. When the Chiricahua Apache band dispersed late in the nineteenth century, some of its members joined their ethnic cousins at the Alamo Chapter of the Navajo Nation, northwest of Magdalena, the community nearest the mountains.

No one seems to know why the range bears the name San Mateo, though it's a common enough Hispanic place name in New Mexico. A greater toponymic mystery is the identity of Withington. His name is on the highest summit in the San Mateos' northern section, as well as on the Withington Wilderness. But all attempts to identify him have failed. It has been suggested that the mountain might have been named by Maj. George M. Wheeler (see *Wheeler Peak*); if so, the most likely eponym would have been William Herbert Withington (1835–1903), Union officer in the Civil War, later Michigan manufacturer and capitalist.

Many mysteries exist in the San Mateos. When Confederate forces retreated southward after the Battle of Glorieta Pass in 1862, they avoided the Rio Grande, and their route may have taken them through the San Mateos' foothills—but if so, where?

Geologically and ecologically, the San Mateos are similar to the other ranges of the Datil-Mogollon Section of the Transition Zone. The eruptions that formed these volcanic mountains began 28–24 million years ago, and the mountains are the remains of three contiguous or overlapping calderas. The northernmost one, centered on Mount Withington, is 27.4 million years old. Adjacent is a smaller one 24.3 million years old. And the one encompassing Vicks Peak is 28.1 million years old. More recent volcanic fields elsewhere in New Mexico are highly basaltic, and their rocks are dark to black, as we expect of volcanic rocks. Here, however, the magma was more silicic, and therefore much lighter in hue, gray or purplish-gray. The rocks have a distinctly angular cleavage as well; they make hiking in lower Potato Canyon like walking over billions of broken floor tiles.

Later, during the Pleistocene, rocks in boulder fields on the steep slopes of the San Mateos became fused together by interstitial ice. This mass flowed downslope just like an ice glacier. With melting of the ice, the rock glaciers remained, suspended in time, as it were. The rock glacier in Shipman Canyon is especially impressive.

Wildlife is abundant in the San Mateo Mountains, as it usually was in Aldo Leopold's time, and for the same reasons: extensive, productive habitat, and few humans. As Leopold wrote:

> The Datil Forest, from the sportsman's standpoint, is the cream
> of the Southwest, and there is no part of the Datil country as
> interesting to hunt in as the San Mateos. It is a very rough region
> with a fair amount of water, and endless miles of yellow
> [ponderosa] pine forest interspersed with oak and piñon.

Myers Cabin, San Mateo Mountains

Any year that the oak fails to produce acorns it is pretty
certain that the piñons will produce nuts, so that the game
is nearly always "hog-rolling fat." (Brown and Carmony,
Aldo Leopold's Southwest)

Two Forest Service wilderness areas are here: the Withington Wilderness
in the north, the Apache Kid Wilderness in the south. But ecologically they
are one unit, as the Forest Service had intended until the objections of a local
rancher and the presence of Forest Road 330 forced their meiosis. Combined,
their total of 63,512 acres would comprise New Mexico's fourth largest wilderness area, though still less than a third the size of the 202,016-acre Aldo
Leopold Wilderness just to the south in the neighboring Black Range.

An extensive system of trails knits the mountains together, though
many are rough and difficult to follow. In the north, in the Withington
Wilderness, the only trails are in Water and Potato Canyons and the link
connecting them. Potato Canyon is especially interesting as it leads to
monoliths resembling giant potatoes.

In the south, most hikers do the 4.2-mile one-way hike from Springtime
Campground to 10,139-foot San Mateo Peak (not to be confused with San
Mateo Mountain), via the Apache Kid Trail, No. 43, and the Cowboy Trail,
No. 44. The summit's fire tower is a worthy destination. Also, from Forest
Road 225 leading to Springtime Campground a trail leads north into the
cliffs overlooking Indian Creek.

In the southwestern San Mateos, the Shipman Trail, reached by a long
and often difficult drive, leads to Myers Cabin via a three-mile hike that
ascends the gully between Vicks Peak and San Mateo Mountain, gaining
about two thousand feet. But none of the trails, even the one to San Mateo
Peak, is frequently traveled. The San Mateo Mountains still offer refuge to
persons seeking solitude, just as they did for the Apache Kid.

Elevations List

The following list is based upon the elevations in the U.S. Geological Survey's USGS 7.5-minute quadrangles and upon the USGS Geographic Names Information System database. It does not include unnamed summits, which in some instances are the highest points in their ranges.

Mountain	Elevation (feet/meters)	Range or region
Wheeler Peak	13,161/4,011	Taos Mountains
Old Mike Peak	13,113/3,997	Taos Mountains
Truchas Peak	13,102/3,993	Santa Fe Mountains
Middle Truchas Peak	13,066/3,984	Santa Fe Mountains
North Truchas Peak	13,024/3,970	Santa Fe Mountains
Simpson Peak	12,976/3,955	Taos Mountains
Lake Fork Peak	12,881/3,926	Taos Mountains
State Line Peak	12,867/3,922	Culebra Range
Chimayosos Peak	12,841/3,914	Santa Fe Mountains
Jicarita Peak	12,835/3,912	Taos Mountains
Big Costilla Peak	12,739/3,883	Taos Mountains
Venado Peak	12,734/3,881	Taos Mountains
Gold Hill	12,716/3,876	Taos Mountains
Latir Peak	12,708/3,873	Taos Mountains
Latir Mesa	12,692/3,869	Taos Mountains
Red Dome	12,688/3,865	Taos Mountains

Mountain	Elevation (feet/meters)	Range or region
Vallecito Mountain	12,643/3,854	Taos Mountains
Santa Fe Baldy	12,622/3,847	Santa Fe Mountains
Sheepshead	12,600/3,840	Taos Mountains
Virsylvia Peak	12,594/3,839	Taos Mountains
Little Costilla Peak	12,584/3,836	Taos Mountains
East Pecos Baldy	12,529/3,819	Santa Fe Mountains
Santa Barbara	12,515/3,815	Santa Fe Mountains
(unofficial name, just east of Chimayosos Peak)		
West Pecos Baldy	12,500/3,810	Santa Fe Mountains
Jicarilla Peak	12,494/3,808	Santa Fe Mountains
Lew Wallace Peak	12,449/3,794	Taos Mountains
Cabresto Peak	12,448/3,794	Taos Mountains
Baldy Mountain	12,441/3,792	Cimarron Range
Lake Peak	12,409/3,782	Santa Fe Mountains
Redonda Peak	12,357/3,766	Santa Fe Mountains
Little Jicarita Peak	12,328/3,758	Santa Fe Mountains
Pueblo Peak	12,305/3,751	Taos Mountains
Penitente Peak	12,249/3,733	Santa Fe Mountains
Taos Cone	12,227/3,727	Taos Mountains
Capulin Peak	12,200/3,719	Santa Fe Mountains
Trampas Peak	12,170/3,709	Santa Fe Mountains
Fraser Mountain	12,163/3,707	Taos Mountains
Lobo Peak	12,115/3,693	Taos Mountains
Baldy Mountain	12,046/3,672	Taos Mountains
Touch-Me-Not Mountain	12,045/3,671	Cimarron Range
Tesuque Peak	12,043/3,671	Santa Fe Mountains
Larkspur Peak	11,992/3,655	Taos Mountains
South Fork Peak	11,978/3,651	Taos Mountains
Sierra Blanca	11,973/3,649	White Mountains
Pinabete Peak	11,948/3,642	Taos Mountains
Flag Mountain	11,946/3,641	Taos Mountains
Cerro Vista	11,939/3,639	Santa Fe Mountains

Mountain	Elevation (feet/meters)	Range or region
Cerro de laOlla	11,932/3,637	Isolated—north-central New Mexico
Sierra Mosca	11,801/3,597	Santa Fe Mountains
Ripley Point	11,799/3,596	Santa Fe Mountains
Relica Peak	11,784/3,592	Taos Mountains
Perra Peak	11,765/3,586	Taos Mountains
Bunkhouse Bare Point	11,730/3,575	Cimarron Range
Mount Phillips	11,696/3,565	Cimarron Range
Tunnel Hill	11,668/3,556	Taos Mountains
Elk Mountain	11,661/3,554	Santa Fe Mountains
Bull-of-the-Woods Mountain	11,640/3,547	Taos Mountains
Vermejo Peak	11,610/3,539	Cimarron Range
Chicoma Mountain	11,561/3,524	Jemez Mountains
Tolby Peak	11,527/3,513	Cimarron Range
Lookout Mountain	11,496/3,504	White Mountains
Mount Barker	11,445/3,488	Santa Fe Mountains
Grouse Mesa	11,403/3,476	Tusas Mountains
Cuchillo de Fernando	11,395/3,473	Rincon Mountains
The Dome	11,336/3,455	Santa Fe Mountains
Black Mountain	11,302/3,445	Taos Mountains
Mount Taylor	11,301/3,445	Mount Taylor
Comanche Peak	11,299/3,444	Cimarron Range
Brazos Peak	11,288/3,441	Tusas Mountains
Taos Peak	11,257/3,431	Taos Mountains
Cerro del Oso	11,255/3,431	Santa Fe Mountains
Redondo Peak	11,254/3,430	Jemez Mountains
Greenie Peak	11,249/3,429	Taos Mountains
Polvadera Peak	11,232/3,424	Jemez Mountains
Van Diest Peak	11,223/3,421	Taos Mountains
Ortiz Peak	11,209/3,417	Taos Mountains
Spring Mountain	11,180/3,408	Santa Fe Mountains

Mountain	Elevation (feet/meters)	Range or region
Aspen Peak	11,109/3,386	Santa Fe Mountains
Agua Fria Peak	11,086/3,379	Cimarron Range
Angel Fire Mountain	11,060/3,371	Cimarron Range
La Mosca Peak	11,053/3,369	Mount Taylor
Peñasco Mountain	10,979/3,346	Santa Fe Mountains
Sawmill Mountain	10,962/3,341	Taos Mountains
Cerro Toledo	10,925/3,330	Jemez Mountains
Garcia Peak	10,925/3,330	Cimarron Range
Canjilon Mountain	10,913/3,326	Tusas Mountains
San Antonio Mountain	10,908/3,325	Isolated—north-central New Mexico
Whitewater Baldy	10,895/3,321	Mogollon Mountains
Black Mountain	10,892/3,320	Cimarron Range
Gallina Mountain	10,887/3,318	Taos Mountains
Osha Mountain	10,885/3,318	Santa Fe Mountains
Cerrito del Padre	10,875/3,315	Santa Fe Mountains
Lucero Peak	10,831/3,301	Taos Mountains
Round Mountain	10,809/3,295	Santa Fe Mountains
Picuris Peak	10,801/3,292	Picuris Mountains
South Baldy	10,783/3,287	Magdalena Mountains
Tetilla Peak	10,782/3,286	Culebra Range
Mogollon Baldy Peak	10,770/3,283	Mogollon Mountains
Buck Mountain	10,748/3,276	White Mountains
Peñasco Amarillo	10,712/3,265	Tusas Mountains
Jawbone Mountain	10,680/3,255	Tusas Mountains
Sandia Crest	10,678/3,255	Sandia Mountains
La Cueva Peak	10,669/3,252	Santa Fe Mountains
Bear Mountain	10,663/3,250	Cimarron Range
Sacaton Peak	10,658/3,249	Mogollon Mountains
Black Mountain	10,643/3,244	Mogollon Mountains
Cerro Saragate	10,641/3,243	Tusas Mountains
The Knob	10,625/3,238	Taos Mountains
Bonito Peak	10,616/3,235	Cimarron Range

Mountain	Elevation (feet/meters)	Range or region
Cieneguilla Mountain	10,613/3,235	Cimarron Range
Cerro de la Garita	10,601/3,231	Jemez Mountains
Pyramid Peak	10,597/3,230	Santa Fe Mountains
San Pedro Peak	10,592/3,228	Sierra Nacimiento
Thompson Peak	10,554/3,217	Santa Fe Mountains
Center Baldy	10,535/3,211	Mogollon Mountains
Gallegos Peak	10,528/3,209	Santa Fe Mountains
Sugar Loaf Mountain	10,525/3,208	Tusas Mountains
Timber Mountain	10,510/3,203	Magdalena Mountains
Rosilla Peak	10,500/3,200	Santa Fe Mountains
Caballo Mountain	10,496/3,199	Jemez Mountains
Hamilton Mesa	10,483/3,195	Santa Fe Mountains
Capulin Peak	10,470/3,191	Taos Mountains
Cimarroncito Peak	10,468/3,191	Cimarron Range
Pajarito Mountain	10,441/3,182	Jemez Mountains
Cerro Rubio	10,449/3,185	Jemez Mountains
Cerros de los Posos	10,449/3,185	Jemez Mountains
North Sandia Peak	10,447/3,184	Sandia Mountains
Olguin Mesa	10,436/3,181	Tusas Mountains
Cerro del Grant	10,432/3,180	Jemez Mountains
Banco Julian	10,413/3,174	Tusas Mountains
Black Mountain	10,390/3,167	Cimarron Range
Shell Mountain	10,385/3,165	Jemez Mountains
Ocate Mesa	10,379/3,164	Cimarron Range
Sierra de Don Fernando	10,365/3,159	Taos Mountains
Broke Off Mountain	10,357/3,157	Tusas Mountains
West Blue Mountain	10,336/3,150	San Mateo Mountains
Cerros del Abrigo	10,332/3,149	Jemez Mountains
East Blue Mountain	10,319/3,145	San Mateo Mountains
Spruce Hill	10,313/3,143	Cimarron Range
Cerro Pavo	10,310/3,142	Jemez Mountains
Cerro Picacho	10,290/3,136	Santa Fe Mountains
Hermit Peak	10,260/3,127	Santa Fe Mountains

Mountain	Elevation (feet/meters)	Range or region
Bear Mountain	10,253/3,125	Santa Fe Mountains
Cerrito Colorado	10,246/3,123	Fernando Mountains
Vicks Peak	10,252/3,125	San Mateo Mountains
Alegres Mountain	10,244/3,122	Isolated
Trail Peak	10,242/3,122	Cimarron Range
Indian Point	10,241/3,121	Jemez Mountains
Casita Piedra	10,210/3,112	Taos Mountains
Apache Peak	10,204/3,110	Taos Mountains
Glorieta Baldy	10,201/3,109	Santa Fe Mountains
Cerro Grande	10,199/3,109	Jemez Mountains
Burned Mountain	10,189/3,105	Tusas Mountains
McKnight Mountain	10,165/3,098	Black Range
Chaperito Knob	10,163/3,098	Santa Fe Mountains
San Mateo Mountain	10,145/3,092	San Mateo Mountains
Tusas Mountain	10,143/3,092	Tusas Mountains
San Mateo Peak	10,139/3,090	San Mateo Mountains
Grouse Mountain	10,135/3,089	Mogollon Mountains
Los Griegos	10,117/3,084	Jemez Mountains
Indian Peak	10,115/3,083	Mogollon Mountains
Mount Withington	10,115/3,083	San Mateo Mountains
Palo Encebado	10,165/3,098	Taos Mountains
Cerro Pelado	10,109/3,081	Jemez Mountains
Manzano Peak	10,098/3,078	Manzano Mountains
Ute Mountain	10,093/3,078	Isolated—north-central New Mexico
Capitan Peak	10,083/3,073	Capitan Mountains
Apache Kid Peak	10,048/3,063	San Mateo Mountains
La Grulla Ridge	10,032/3,058	Cimarron Range
Reeds Peak	10,015/3,053	Black Range
Hillsboro Peak	10,011/3,051	Black Range
Gallo Peak	10,003/3,049	Manzano Mountains
Mogote Peak	9,9875/3,043	Tusas Mountains
San Antonio Mountain	9,978/3,041	Jemez Mountains

Mountain	Elevation (feet/meters)	Range or region
Iron Mountain	9,965/3,037	Cimarron Range
Nogal Peak	9,957/3,035	White Mountains
Teepee Peak	9,957/3,035	San Mateo Mountains
Bearwallow Mountain	9,953/3,033	Mogollon Mountains
La Mesa del Pedregosa	9,949/3,032	Jemez Mountains
Tower Peak	9,949/3,032	San Mateo Mountains
Burn Peak	9,938/3,029	Cimarron Range
Corner Mountain	9,938/3,029	Mogollon Mountains
Rabbit Mountain	9,938/3,029	Jemez Mountains
Las Conchas	9,934/3,028	Jemez Mountains
El Cielo Mountain	9,930/3,026	Santa Fe Mountains
Lookout Peak	9,922/3,024	Cimarron Range
Chromo Mountain	9,916/2,022	Extension of Colorado's San Juan Mountains
Cerro Jarocito	9,886/3,013	Sierra Nacimiento
Cerro del Medio	9,881/3,012	Jemez Mountains
Apache Peak	9,872/3,008	Cimarron Range
Escondido Mountain	9,869/3,008	Mangas Mountains
Cerro Pedernal	9,862/3,006	Jemez Mountains
Viveash Mesa	9,862/3,006	Santa Fe Mountains
North Baldy	9,858/3,005	Magdalena Mountains
Cerro Pelon	9,852/3,003	Jemez Mountains
Diamond Peak	9,850/3,002	Black Range
Grass Mountain	9,841/3,000	Santa Fe Mountains
Quartzite Peak	9,829/2,996	Tusas Mountains
Windy Point	9,820/2,993	Mogollon Mountains
Rayado Peak	9,805/2,989	Cimarron Range
Nacimiento Peak	9,801/2,987	Sierra Nacimiento
Elk Mountain	9,799/2,987	Elk Mountains
Eagle Peak	9,786/2,983	Tularosa Mountains
Cuchillo del Medio	9,785/2,982	Taos Mountains
West Baldy	9,785/2,982	Mogollon Mountains
South Sandia Peak	9,782/2,982	Sandia Mountains

Mountain	Elevation (feet/meters)	Range or region
Holt Mountain	9,780/2,981	Mogollon Mountains
Sheep Mountain	9,779/2,981	Black Range
Turkeyfeather Mountain	9,771/2,978	Mogollon Mountains
Scully Mountain	9,765/2,976	Cimarron Range
Costilla Peak	9,752/2,972	Culebra Range
Crater Peak	9,748/2,971	Cimarron Range
Lone Pine Mesa	9,744/2,970	Santa Fe Mountains
Kiowa Mountain	9,735/2,967	Tusas Mountains
Mogotito	9,726/2,964	Tusas Mountains
Johnson Mesa	9,722/2,963	Santa Fe Mountains
Cub Mountain	9,710/2,959	Mogollon Mountains
Mesa del Medio	9,706/2,958	Jemez Mountains
Buzzard Peak	9,692/2,954	Tularosa Mountains
Mangas Mountain	9,691/2,954	Mangas Mountains
Spring Mountain	9,683/2,951	Mogollon Mountains
Cerro Osha	9,678/2,950	Mount Taylor
Sawyers Peak	9,668/2,947	Black Range
Cerrito Colorado	9,661/2,945	Taos Mountains
A-L Peak	9,648/2,941	San Mateo Mountains
Aspen Hill	9,646/2,940	North of Ocate
Monjeau Peak	9,641/2,939	White Mountains
Padilla Point	9,627/2,934	Capitan Mountains
Cross-O Mountain	9,620/2,932	Black Range
Mining Mountain	9,617/2,931	Sierra Nacimiento
Bosque Peak	9,610/2,929	Manzano Mountains
Carrizo Mountain	9,605/2,928	Isolated—south-central New Mexico
Aspen Mountain	9,606/2,928	Black Range
Eureka Mesa	9,592/2,924	Sierra Nacimiento
Madre Mountain	9,556/2,913	Datil Mountains
North Bosque Peak	9,549/2,911	Manzano Mountains
Granite Knob	9,545/2,909	White Mountains
Pinoreal	9,524/2,903	Fernando Mountains
Mosca Peak	9,509/2,898	Manzano Mountains

Highest County Summits

Bernalillo
South Sandia Peak	9,782
Cedro Peak	7,767
Mount Washington	7,716

Catron
Whitewater Baldy	10,895
Mogollon Baldy Peak	10,770
Sacaton Mountain	10,658

Chaves
One Tree Peak	7,089
Chimney Peak	7,060
Wind Mountain	6,778

Cibola
Mount Taylor	11,301
Cerro Pelon	10,028
Cerro Osha	9,678

Colfax
Little Costilla Peak	12,584
Baldy Mountain	12,441
Touch-Me-Not Mountain	12,045

Curry
Bonney Hill	4,468
[No others named in GNIS]	

DeBaca
Loma Alta	5,583
El Morro Mesa	5,013
Mesita de Anil	4,925

Doña Ana
Organ Needle	8,990
Organ Peak	8,872
Baldy Peak	8,445

Eddy
Deer Hill	7,056
Wild Cow Mesa	6,500
Pickett Hill	6,450

Grant
McKnight Mountain	10,165
Sawyers Peak	9,668
Cross-O Mountain	9,620

Guadalupe
Cerro de Pedro Miguel	6,862
Mesa Leon	6,339
Potrillo Hill	6,175

Harding
Sugarloaf Mountain	6,455
Spear Hills	6,228
Cerro de la Cruz	6,184

Hidalgo		*Otero*	
Animas Peak	8,482	Sierra Blanca	11,973
Big Hatchet Peak	8,320	Cow Mountain	9,396
Attorney Mountain	8,005	Alamo Peak	9,260

Lea		*Quay*	
Soldier Hill	4,421	Palomas Mesa	5,325
Taylor Peak	4,314	Mesa Redonda	5,192
Hat Mesa	3,930	Sand Mountain	5,170

Lincoln		*Rio Arriba*	
Capitan Peak	10,083	Truchas Peak	13,102
Nogal Peak	9,957	Middle Truchas Peak	13,066
Monjeau Peak	9,641	North Truchas Peak	13,024

Los Alamos		*Roosevelt*	
Guaje Mountain	7,636	The Mesa	4,712
Twomile Mesa	7,490	Negrohead	4,682
North Mesa	7,385	Chalk Hill	4,038

Luna		*San Juan*	
Cookes Peak	8,408	Beautiful Mountain	9,388
Florida Peak	7,448	Little White Cone	8,411
Baldy Peak	7,106	Huerfano Mountain	7,474

McKinley		*San Miguel*	
Cerro Redondo	8,976	Elk Mountain	11,661
Cerros de Alejandro	8,969	Mount Barker	11,445
Cerro Chivato	8,917	Spring Mountain	11,180

Mora		*Sandoval*	
Pecos Baldy	12,500	Redondo Peak	11,254
Round Mountain	10,809	Cerro de la Garita	10,601
Cerrito del Padre	10,785	Caballo Mountain	10,496

Santa Fe

Santa Fe Baldy	12,622
Lake Peak	12,409
Redonda Peak	12,357

Sierra

Reeds Peak	10,015
Hillsboro Peak	10,011
Diamond Peak	9,850

Socorro

South Baldy	10,783
Timber Peak	10,510
West Blue Mountain	10,336

Taos

Wheeler Peak	13,161
Old Mike Peak	13,113
Simpson Peak	12,976

Torrance

Manzano Peak	10,098
Gallo Peak	10,003
Bosque Peak	9,610

Union

Sierra Grande	8,720
Capulin Mountain	8,182
Davis Mesa	7,527

Valencia

Chicken Mountain	7,826
Mesa Lucero	6,850
Cerro del Indio	6,330

Glossary of Selected Geological Terms

Agate: very finely crystalline quartz, often with multi-colored bands making it a semi-precious stone.

Alkali: sodium and potassium carbonates and other salts, including sodium chloride, usually noticed as a white crust, at or near the surface in arid or semiarid areas.

Alluvial fan: a deltalike deposit of mud, sand, and gravel at canyon mouths in arid or semiarid regions.

Andesite: a volcanic rock intermediate in composition between rhyolite and basalt.

Anticline: a bulging upward of the Earth's crust, often but not always linear; a convex fold, like a squeezed blanket.

Bajada: a slope of gravel and sand formed by overlapping alluvial fans.

Basalt: a fine-grained, dark-colored volcanic rock poor in silica but rich in iron and magnesium.

Batholith: a large mass of igneous rock that has solidified deep within the Earth.

Bedrock: the solid rock encountered when digging; solid rock exposed at or near the surface.

Breccia (volcanic): volcanic rock comprised of broken, angular rock fragments embedded in a finer matrix, such as volcanic ash or lava.

Butte: a flat-topped prominence, formed by differential erosion of a hard layer, such as sedimentary strata or volcanic flows, underlain by softer layers. Conspicuous in the arid Southwest, buttes are distinguished from mesas in being taller than they are wide.

Calcite: a common and extremely varied mineral, typically formed by the precipitation and crystallization of calcium carbonate, often as veins in other minerals. Among its many forms are travertine, formed in caves and springs; tufa, soft, white travertine spring deposits; chalk; alabaster; and onyx. Transparent calcite crystals, attractive but very soft, often are found on the ground in old sandstone formations.

Caldera: a basin-shaped depression formed by a volcanic explosion and subsequent collapse of a magma chamber. Valles Caldera, Jemez Mountains.

Capilla: Spanish, "hood, cowl," often used as a descriptive metaphor for mountains.

Caprock: a resistant rock or layer protecting softer layers beneath from erosion.

Cerro: Spanish, "hill," though in New Mexico the term can refer to landforms ranging from small lowland hills to high summits of major mountain ranges. The diminutives are *cerrillo* and *cerrito*.

Cinder cone: a small, conical volcano built up of frothy, "popcorn-like" volcanic material that issued from a single vent. (See *Capulin Mountain*.)

Cirque: a circular valley with precipitous walls, often created through glacial action.

Colorado Plateau: the geologic province located between the Basin and Range Province on the south and east and the Rocky Mountains on the north and west, and including the portions of Colorado, Utah, Arizona, and New Mexico around their mutual meeting at the Four Corners. Characterized by deep, predominantly horizontal strata of sedimentary rocks, interrupted often by volcanic features.

Copper: essential metallic element, occurring as native copper or combined with sulfides or carbonates, typically found in veins or volcanic rocks, such as in southwestern New Mexico, location of the large pit mines at Santa Rita and Tyrone. Copper oxidizes into colorful minerals such as green malachite and chrysocolla and blue azurite.

Crestón: in New Mexico, the Spanish *crestón* can mean "hillock" or "summit," but it also can mean "cockscomb" or "large crest," often referring to sharp ridges capped by rock outcrops.

Cuchillo: Spanish, "knife," often a descriptive metaphor for sharp ridges.

Conglomerate: a sedimentary rock made up of rounded gravel and pebbles cemented together by another mineral.

Cuesta: a linear ridge, one of whose slopes downward from the crest is long and gentle, the other short and steep; imagine a brick tilted on one of its long edges.

Cumbre: Spanish, "summit."

Dacite: a volcanic rock rich in quartz and feldspar.

Diatreme: the throat of a volcano after the surrounding rocks have been eroded away. Ship Rock.

Dike: a sheetlike intrusion of igneous rock into cracks or joints in the surrounding rock.

Dome: an upward bulge in the Earth's crust (see *anticline*), from whose top rocks slope downward in all directions.

Era: the largest unit of geologic time, of which the Earth has had four: Precambrian, Paleozoic, Mesozoic, and Cenozoic.

Epoch: the third-order unit of geological time.

Erosion: the process of materials on the Earth's surface being worn away and carried off, by such agents as water, wind, dissolved chemicals, and others.

Escarpment: a long, more or less continuous, relatively steep slope or face breaking the general continuity of the land surface. West face of the Sandias.

Extrusive: describes igneous rocks that have erupted and cooled at the Earth's surface; all lava formations are extrusive.

Fault: a rock fracture along which movement has occurred.

Fault block: a segment of the Earth's crust bounded on two or more sides by faults.

Fault scarp: a steep slope or cliff formed when rocks are displaced vertically along a fault.

Feldspar: a large group of minerals made of silica and aluminum oxides, light-colored, rock-forming minerals. They make up sixty percent of Earth's crust.

Fluorite: found in both igneous and sedimentary rocks, often with lead, copper, and zinc. The United States is the world's foremost producer of fluorite, the bulk of which comes from four states: Illinois, Kentucky, Colorado, and New Mexico.

Formation: the geologic term for a unique, recognizable, extensive rock unit that is distinct from adjacent rock units.

Gold: a metallic element, found in a variety of forms, from amalgams with other minerals to almost microscopic granules to sheets and wires to water-washed nuggets. Found either in veins or lodes or in sand and gravel deposits called *placers.*

Gneiss: metamorphic rock characterized by bands, usually light and dark, of different minerals.

Graben: a portion of Earth's crust, bounded by two or more faults, that has subsided while rock on the other sides of the faults typically is uplifted. The Plains of San Agustin are an example. See *Horst.*

Granite: a coarse-grained igneous intrusive rock composed of chunky crystals of quartz and feldspar peppered with dark biotite and hornblende.

Gypsum: a common mineral, calcium sulfate, formed usually by evaporation of sea water.

Horst: a block of Earth's crust separated by faults from adjacent and relatively lower blocks.

Hogback: a long, narrow ridge with a sharp crest formed by erosion of layers of resistant rock tilted approximately forty-five degrees.

Igneous rock: rock that has formed from molten magma and then cooled and solidified, either beneath the Earth's surface (intrusive rock) or at the surface (extrusive).

Intrusion, intrusive: both refer to molten rock that penetrated into older rock and then cooled before reaching the surface. Subsequent erosion often exposes these rocks at the surface.

Laccolith: a body of igneous rock intruded between older rock layers, causing the upper layers to arch as in a dome.

Laramide Orogeny: a continent-wide period of mountain-building that began about 70 million years ago and lasted until about 40 million years ago.

Lava: igneous rock created by cooling of molten magma on the Earth's surface or under the sea.

Lava dome: a dome-shaped body of volcanic rock formed from very viscous magma.

Limestone: a sedimentary rock consisting of calcium carbonate, usually formed from the shells of marine animals and calcium-secreting plants.

Loma: Spanish, "hill"; *lomita,* "little hill." In New Mexico, these terms typically refer to landforms significantly smaller, with less relief, than hills denoted by the Spanish term cerro.

Magma: molten rock, usually from deep beneath Earth's surface.

Magma chamber: the reservoir of molten rock, typically a few kilometers beneath the surface, whose ascending lava and gasses feed volcanoes.

Marine: rocks laid down in an ocean environment, rather than a freshwater one of lakes and rivers.

Mesa: Spanish, literally, "table," in geology referring to a flat-topped hill or mountain capped with a resistant rock layer and edged with steep cliffs or slopes, broader than it is tall, larger than a butte and smaller than a plateau, though the form and extent of Southwest landforms termed *mesas* vary enormously.

Metamorphic rocks: rocks that have been altered from their original configuration by great heat and pressure or by chemical changes.

Morro: Spanish, "bluff, headland."

Orogeny: mountain-building, generally involving large-scale faulting, magma intrusion, volcanic activity, and folding of the earth's crust. An orogeny represents a period of geological activity rather than a single event.

Pediment: a gently inclined erosion surface carved in bedrock at the base of a mountain range.

Pelon: Spanish, "bald."

Period: a unit of geologic time, shorter than an era, longer than an epoch.

Picacho: Spanish, "peak, pointed summit."

Pico: Spanish, "peak, pointed summit."

Placer deposit: an alluvial or glacial mineral deposit formed by the mechanical concentration of heavy mineral particles, such as gold, from weathered debris.

Plate: a large block of the Earth's crust, separated from other blocks by mid-ocean ridges, trenches, and collision zones.

Plateau: a flat-topped mountain larger than a mesa, usually capped with a resistant rock layer and edged on two or more sides with cliffs or steep slopes.

Playa: shallow, temporary desert lake, formed after rain and soon evaporating.

Pluton: an igneous mass formed by the solidification of molten magma deep within the Earth.

Pumice: light-colored, frothy volcanic rock, often light enough to float on water.

Quartz: a hard, glassy, rock-forming mineral composed of crystalline silica.

Rhyolite: light-colored, very fine grained volcanic rock, the extrusive equivalent of granite.

Rift: a narrow block of the Earth's crust formed by down-dropping between two more or less parallel fault zones that reach down to the mantle. The Rio Grande Rift.

Rock glacier: a glacier-like tongue of broken rock, usually lubricated by water and ice and moving slowly downslope like an ice glacier. Capitan Mountains.

Sandstone: sedimentary rock formed from sand, usually cemented with calcium carbonate.

Schist: metamorphic rock characterized by bands of flaky minerals, chiefly mica.

Sedimentary rock: formed from particles of other rock transported and deposited by water, wind, or ice.

Shield volcano: a broad, low volcano formed by moderately fluid lava flowing some distance from the vent. Sierra Grande.

Sierra: Spanish, "mountain range," though the term sometimes is applied to individual peaks or mountain ridges.

Strata: layers of rock, usually sedimentary but also including lava flows; singular, is stratum.

Stratovolcano: a cone-shaped volcano built of alternating layers of lava and volcanic ash; sometimes called a *composite volcano*. Mount Taylor.

Syncline: a downward folding of rock layers, like sagging blankets.

Talus: the jumble of rock fragments typically found at the base of cliffs or steep slopes, from which they've fallen.

Tent rocks: the teepee-shaped formations formed by erosion in volcanic areas when a rock fragment perches atop a cone of softer rock, which it protects.

Tuff: a rock formed of compacted volcanic ash and cinders.

Turquoise: a copper-and-aluminum phosphate mineral that in its natural state most often appears as white chalk. Highly prized by Native Americans both for its beauty when treated and polished and also its religious significance, often symbolizing the sky. High quality natural turquoise is rare, and the turquoise in most jewelry has been altered to enhance color and strength, and much of it is just plain fake.

Vein: typically a small intrusion into surrounding rock, like roots into soil, often associated with mineral deposits.

Weathering: general physical and chemical processes whereby rocks exposed at the Earth's surface are worn down and decomposed.

Zinc: most often associated with lead and copper ores in igneous rocks or as replacement ores in carbonate rocks. Smithsonite and sphalerite are the important zinc minerals in New Mexico.

Mountain Headings of New Mexico Rivers

Animas River	San Juan Mountains—Colorado
Rio Bonito	White Mountains
Black River	Guadalupe Mountains
Rio Brazos	Tusas Mountains
Canadian River	Raton Mountains
Rio Capulin (Rio Arriba)	Jemez Mountains
Rio Capulin	Sangre de Cristo Mountains—Santa Fe Range
Rio Cebolla	Jemez Mountains
Chama River	San Juan Mountains—Colorado
Rio Chiquito	Sangre de Cristo Mountains—Fernando Mountains
Rio Chiquito	Sangre de Cristo Mountains—Truchas
Rio Chupadero	Sangre de Cristo Mountains—Santa Fe
Cimarron River	Sangre de Cristo Mountains—Cimarron Range
Rio Costilla	Sangre de Cristo Mountains—Culebra Range
Rio de la Cebolla	Sangre de Cristo Mountains—Truchas Range
Rio de la Olla	Sangre de Cristo Mountains—Fernando
Rio de las Trampas	Sangre de Cristo Mountains—Truchas Range
Rio de las Vacas	Sangre de Cristo Mountains—Sierra Nacimiento
Rio de los Pinos	San Juan Mountains—Colorado
Rio de Tierra Amarilla	Brazos Mountains

Rio de Truchas	Sangre de Cristo Mountains—Truchas Range
Rio del Oso	Jemez Mountains
Rio del Plano	Chico Hills
Dry Cimarron River	Davis Mesa
Rio en Medio	Sangre de Cristo Mountains— Santa Fe Mountains
Rio Feliz	White Mountains
Rio Fernando de Taos	Sangre de Cristo Mountains—Taos Range
Rio Frijoles	Jemez Mountains
Rio Frijoles	Sangre de Cristo Mountains— Santa Fe Mountains
Rio Gallina (near Gallina)	Sierra Nacimiento/San Pedro Mountains
Gallinas River	Sangre de Cristo Mountains—Las Vegas Range
Gila River	Mogollon Mountains—Black Range
Rio Grande	San Juan Mountains—Colorado
Rio Grande del Rancho	Sangre de Cristo—Fernando Mountains
Rio Guadalupe	Jemez Mountains
Rio Hondo	White Mountains
Rio Hondo	Sangre de Cristo Mountains—Taos Range
Jemez River	Jemez Mountains
Rio la Casa	Sangre de Cristo Mountains—Las Vegas Range
La Plata River	La Plata Mountains—Colorado
Los Pinos River (McKinley)	San Juan Mountains—Colorado
Rio Lucero	Sangre de Cristo Mountains—Taos Range
Mancos River	La Plata Mountains—Colorado
Rio en Medio	Sangre de Cristo Mountains— Santa Fe Mountains
Mimbres River	Black Range

Mora River	Sangre de Cristo Mountains—Santa Fe Mountains
Rio Mora	Sangre de Cristo Mountains—Santa Fe Mountains
Rio Nambe	Sangre de Cristo Mountains—Santa Fe Mountains
Navajo River	San Juan Mountains—Colorado
Rio Nutria	Zuni Mountains
Rio Nutrias	Brazos Mountains
Rio Ojo Caliente	Tusas Mountains
Rio Peñasco	Sacramento Mountains
Rio Pescado	Zuni Mountains
Pecos River	Sangre de Cristo Mountains—Santa Fe
Rio Pueblo	Sangre de Cristo Mountains—Santa Fe
Rio Pueblo de Taos	Sangre de Cristo Mountains—Taos Range
Rio Puerco	Sierra Nacimiento
Rio Puerco	Jemez Mountains
Puerco River	W of the Continental Divide, at Borrego Pass
Rio Quemado	Sangre de Cristo Mountains—Santa Fe Mountains
Red River	Sangre de Cristo Mountains—Taos Range
Rio Ruidoso	White Mountains
Rio Salado	Jemez Mountains
Rio Salado	Catron County
Rio San Antonio	Tusas Mountains
San Juan River	San Juan Mountains—Colorado
Rio San Leonardo	Sangre de Cristo Mountains—Truchas Range
Rio Santa Barbara	Sangre de Cristo Mountains—Santa Fe Mountains
Santa Fe River	Sangre de Cristo Mountains—Santa Fe Mountains

Rio Tesuque	Sangre de Cristo Mountains— Santa Fe Mountains
Tularosa River	Tularosa Mountains
Rio Tusas	Tusas Mountains
Rio Valdez	Sangre de Cristo Mountains— Santa Fe Mountains
Rio Vallecitos, Upper and Lower	Tusas Mountains
Zuni River	Zuni Mountains

Selected Bibliography

Allen, John Eliot, and Frank E. Kottlowski. "Roswell-Capitan-Ruidoso and Bottomless Lakes State Park, New Mexico," *Scenic Trips to the Geologic Past*. No. 3. Socorro: NM Bureau of Mines and Mineral Resources, 1958.

Atkinson, William W., Jr. *Geology of the San Pedro Mountains, Santa Fe County, NM.* Socorro: Bureau of Mines and Mineral Resources, 1961.

Baldwin, Brewster, and William R. Muehlberger. *Geologic Studies of Union County, New Mexico.* Bulletin 63. Socorro: State Bureau of Mines and Mineral Resources, 1959.

Baars, Donald L. *The Colorado Plateau: a Geologic History.* Albuquerque: University of New Mexico Press, 1983.

———. *Navajo Country: a geology and natural history of the Four Corners Region.* Albuquerque: University of New Mexico Press, 1995.

Barker, Elliott S. *Beatty's Cabin.* Santa Fe: William Gannon, 1977.

Bauer, Paul W. [et al.]. "The Enchanted Circle: loop drives from Taos." *Scenic Trips to the Geologic Past*, No. 2. Socorro: NM Bureau of Mines and Mineral Resources, 1991.

Bauer, Paul W., Richard P. Lozinsky, Carol J. Condie, and L. Greer Price. *Albuquerque: a Guide to its Geology and Culture.* Socorro: NM Bureau of Geology and Mineral Resources, 2003.

Brown, David E. and Neil B. Carmony, eds. *Aldo Leopold's Southwest.* Albuquerque: University of New Mexico Press, 1995.

Chilton, Lance, et al. *New Mexico: a new guide to the colorful state.* Albuquerque: University of New Mexico Press, 1984.

Christiansen, Paige W. "The Story of Mining in New Mexico," *Scenic Trips to the Geologic Past* No. 12. Socorro: NM Bureau of Mines and Mineral Resources, 1974.

Christiansen, Paige W., and Frank E. Kottlowski. "Mosaic of New Mexico's Scenery, Rocks, and History," *Scenic Trips to the Geologic Past.* No. 8. Socorro: NM Bureau of Mines and Mineral Resources, 1972.

Chronic, Halka and Lucy Chronic. *Pages of Stone: Geology of Western National Parks and Monuments.* Seattle: The Mountaineers Books, 2004.

———. *Roadside Geology of New Mexico,* Missoula, MT: Mountain Press Publishing, 1987.

Clark, Kenneth F., and Charles B. Read. *Geology and Ore Deposits of Eagle Nest Area, New Mexico.* Bulletin 94. Socorro: State Bureau of Mines and Mineral Resources, 1972.

Clemons, Russell E. "A Trip through Space and Time: Las Cruces to Cloudcroft," *Scenic Trips to the Geologic Past* No. 15. Socorro: NM Bureau of Mines and Mineral Resources, Socorro, NM, 1996.

Clemons, R. E. *Geology of Good Sight Mountains and Uvas Valley, Southwest New Mexico.* Circular 169. Socorro: State Bureau of Mines and Mineral Resources, 1979.

Couchman, Donald Harold. *Cookes Peak—Pasaron por Aqui: a Focus on US History in Southwest New Mexico.* Bureau of Land Management—Mimbres Resource Area, 1990.

Crumpler, L.S. and J.C. Aubele. *Volcanoes of New Mexico: an Abbreviated Guide for Non-specialists,* NM Museum of Natural History and Science, Bulletin 18 (2001): 5–15.

Deal, Edmond G. *Geology of the Northern Part of the San Mateo Mountains.* Ph.D. diss. University of New Mexico, 1973.

DeBuys, William. *Enchantment and Exploitation: the Life and Hard Times of a New Mexico Mountain Range.* Albuquerque: University of New Mexico Press, 1985.

Degenhardt, William G., Charles W. Painter, and Andrew H. Price. *Amphibians and Reptiles of New Mexico.* Albuquerque: University of New Mexico Press, 1996.

Dick-Peddie, William A. *New Mexico Vegetation: Past, Present, and Future.* Albuquerque: University of New Mexico Press, 1993.

Disbrow, Alan E., and Walter C. Stoll. *Geology of the Cerrillos Area.* Socorro: Bureau of Mines and Mineral Resources, 1957.

Dunham, Kingsley Charles. *The Geology of the Organ Mountains.* Bulletin 11. Socorro: NM Bureau of Mines and Mineral Resources, 1935.

Ebricht, Malcom. *Land Grants and Lawsuits in Northern New Mexico.* Albuquerque: University of New Mexico Press, 1994.

Farkas, Steven Eugene. *Geology of the Southern San Mateo Mountains.* Ph.D. diss. Unversity of New Mexico, 1968.

Findley, James S., Arthur H. Harris, Don E. Wilson, and Clyde Jones. *Mammals of New Mexico.* Albuquerque: University of New Mexico Press, 1968.

Fish, Jim, ed. *Wildlands: New Mexico BLM Wilderness Coalition Statewide Proposal,* 1987.

Foster, Roy W. *Southern Zuni Mountains.* Socorro: Bureau of Mines and Mineral Resources, 1975.

Fugate, Francis L., and Roberta B Fugate. *Roadside History of New Mexico.* Missoula, MT: Mountain Press Publishing Co., 1989.

Gillerman, Elliot. *Geology of the Central Peloncillo Mountains, Hidalgo County, NM, and Cochise County, AZ.* Bulletin 57. Socorro: State Bureau of Mines and Mineral Resources, 1958.

Glover, Vernon J. *Jemez Mountains Railroads: Santa Fe National Forest.* Historical Society of New Mexico, 1990. 917.8957 G566 Alb Pub E Mtn.

Harley, George Townsend. *The Geology and Ore Deposits of Sierra County, New Mexico.* Bulletin 10. Socorro: NM Bureau of Mines and Mineral Resources, 1934.

Hewett, Edgar L. *Landmarks of New Mexico.* Albuquerque: University of New Mexico Press, 1947.

Hill, Mike. *Guide to the Hiking Areas of New Mexico.* Albuquerque: University of New Mexico Press, 1995.

———. *Hikers and Climbers Guide to the Sandias.* 2nd ed. Albuquerque: University of New Mexico Press, 1993.

Hoard, Dorothy. *A Guide to Bandelier National Monument.* Los Alamos: Los Alamos Historical Society, 1983.

———. *Los Alamos Outdoors.* 2nd. ed. Los Alamos: Los Alamos Historical Society, 1995.

Howe, Wesley M. *From Basin to Peak: an Explorer's Guide to the Colorado-New Mexico San Juan Basin.* Lubbock, TX: Texas Tech University Press, 1998.

Hubler, Clark. *America's Mountains: an Exploration of Their Origins and Influence.* Facts on File, 1995.

Ivey, Robert DeWitt. *Flowering Plants of New Mexico.* Albuquerque: self-published, 1995.

Julyan, Bob. *New Mexico's Continental Divide Trail: the Official Guide.* Englewood, CO: Westcliffe Publishers, 2000.

———. *New Mexico's Wilderness Areas: the Complete Guide.* Englewood, CO: Westcliffe Publishers, 1999.

Julyan, Robert. *The Place Names of New Mexico.* Albuquerque: University of New Mexico Press, 1996.

Julyan, Robert and Mary Stuever, eds. *Field Guide to the Sandia Mountains.* Albuquerque: University of New Mexico Press, 2005.

Kelley, Vincent C., and Stuart A. Northrop. *Geology of Sandia Mountains and Vicinity.* Memoir 29. Socorro: NM Bureau of Geology and Mineral Resources, 1975.

Kues, Barry S. *The Fossils of New Mexico.* Albuquerque: University of New Mexico Press, 1982.

Larson, Peggy. *The Deserts of the Southwest.* San Francisco: Sierra Club Books, 1977.

Lasky, Samuel G. *The Ore Deposits of Socorro County.* Socorro: Bureau of Mines and Mineral Resources, 1932.

Logan, Kenneth A. and Linda L. Sweanor. *Desert Puma: Evolutionary Ecology and Conservation of and Enduring Carnivore.* Washington, D.C.: Island Press, 2001.

Lozinsky, Richard P., Richard W. Harris, and Stephen Leksohn. *Elephant Butte—Eastern Black Range Region.* Socorro: NM State Bureau of Mines and Mineral Resources, 1995.

McDonald, Corry. *Wilderness: a New Mexico Legacy.* Santa Fe: Sunstone Press, 1985.

McKenna, James. *Black Range Tales: Chronicling Sixty Years of Life and Adventure in the Southwest.* Glorieta, NM: Rio Grande Press, 1936.

McPhee, John. *Basin and Range.* New York: Farrar, Straus, and Giroux, 1981.

MacCarter, Jane S. *New Mexico Wildlife Viewing Guide.* Helena, MT: Falcon, 2000.

MacDonald, Jerry. *Earth's First Steps: Tracking Life Before the Dinosaurs.* Boulder, CO: Johnson Books, 1994.

Mack, Greg H. *The Geology of Southern New Mexico.* Albuquerque: University of New Mexico Press, 1977. Santa Fe Public, SW 557.89 Mac.

Magee, Greg S. *A Hiking Guide to Doña Ana County.* Las Cruces: Roseta Press, 1989.

Mangan, Frank. *Ruidoso Country.* El Paso, TX: Mangan Books, 1994.

Matthews, Kay. *Hiking the Mountain Trails of Santa Fe.* Chamisal, NM: Acequia Madre Press, 1995.

————. *Hiking the Wilderness: a Backpacking Guide to the Wheeler Peak, Pecos, and San Pedro Parks Wilderness Areas.* Chamisal, NM: Acequia Madre Press, 1992.

Meriwether, Frank. *My Life in the Mountains and on the Plains: the Newly Discovered Autobiography.* Robert A. Griffen, ed. Norman, OK: University of Oklahoma Press, 1965.

Miller, John P., Arthur Montgomery, and Patrick K. Sutherland. *Geology of Part of the Southern Sangre de Cristo Mountains.* Socorro: Bureau of Mines and Mineral Resources, 1963.

Mitchell, James R. *Gem Trails of New Mexico.* Baldwin Park, CA: Gem Guides Book Company, 2001.

Montgomery, Arthur. *Pre-Cambrian Geology of the Picuris Range, North-Central New Mexico.* Bulletin 30. Socorro: State Bureau of Mines and Mineral Resources, 1953.

Morris, Irvin. *From the Glittering World: a Navajo Story.* Norman, OK: University of Oklahoma Press, 1997.

Murphy, Dan. *The Guadalupes: Guadalupe Mountains National Park.* Carlsbad, NM: Carlsbad Caverns Natural History Association, 1984.

Murray, John. *The Gila Wilderness: a Hiking Guide.* Albuquerque: University of New Mexico Press, 1988.

New Mexico Geological Society Field Conference. *Guidebook of South-Central New Mexico.* Sixth Field Conference, November 11–12 & 13, 1955. Socorro, N.M.: The Society, 1955.

New Mexico Geological Society Field Conference. *Guidebook of Santa Fe Country.* Raymond V. Ingersoll, editor. Thirtieth Field Conference, October 4–6, 1979. Socorro, N.M.: The Society, 1979.

Northrup, Stuart A. *Minerals of New Mexico.* 3rd ed. revised by Florence A. LaBruzza. Albuquerque: University of New Mexico Press, 1996.

Pearson, Jim Berry. *The Red River—Twining Area: a New Mexico Mining Story.* Albuquerque: University of New Mexico Press, 1986.

Perhace, Ralph M. *Geology and Mineral Deposits of the Gallinas Mountains, Lincoln and Torrance Counties, New Mexico.* Bulletin 95. Socorro: State Bureau of Mines and Mineral Resources, 1970.

Price, L. Greer. *An Introduction to Grand Canyon Geology.* Grand Canyon, AZ: Grand Canyon Association, 1999.

Pynes, Patrick. "Chuska Mountains and Defiance Plateau, Navajo Nation," Essay on Colorado Plateau, Land Use History of North America (CP-LUHNA) on internet. http://www.cpluhna.nau.edu/Places/chuska_mtns.htm.

Robinson, Sherry. *El Malpais, Mt. Taylor, and the Zuni Mountains: a Hiking Guide and History.* Albuquerque: University of New Mexico Press, 1994.

Sandersier, Andy. *The Lakes of New Mexico: a Guide to Recreation.* Albuquerque: University of New Mexico Press, 1996.

Schmidt, Paul G., and Campbell Craddock. *The Geology of the Jarilla Mountains, Otero County, NM.* Bulletin 82. Socorro: State Bureau of Mines and Mineral Resources, 1964.

Seager, William R., John W. Hawley, and Russell E. Clemons. *Geology of San Diego Mountain Area, Doña Ana County, NM.* Bulletin 97. Socorro: State Bureau of Mines and Mineral Resources, 1971.

Seager, William R., Frank E. Kottlowski, and John W. Hawley. *Geology of Doña Ana Mountains, New Mexico.* Circular 147. Socorro: State Bureau of Mines and Mineral Resources, 1976.

Seager, William R., and Greg H. Mack. *Geology of East Potrillo Mountains and Vicinity, Doña Ana County, NM.* Bulletin 113. Socorro: State Bureau of Mines and Mineral Resources, 1994.

Sherman, James E. and Barbara H Sherman. *Ghost Towns and Mining Camps of New Mexico.* Norman, OK: University of Oklahoma Press, 1975.

Simmons, Marc. *New Mexico: An Interpretive History.* Albuquerque: University of New Mexico Press, 1988.

Sonnichsen, C.L. *Tularosa: Last of the Frontier West.* Albuquerque: University of New Mexico Press, 1960.

Soule, J.M. *Structural Geology of Northern Part of Animas Mountains, Hidalgo County, New Mexico.* Circular 125. Socorro: State Bureau of Mines and Mineral Resources, 1972.

Sutherland, Patrick K., and Arthur Montgomery. *Trail Guide to the Geology of the Upper Pecos.* Socorro: NM Bureau of Mines and Mineral Resources, 1975.

Thornton, Major William Anderson. *Diary of William Anderson Thornton: Military Expedition to New Mexico.* The Kansas Collection.

Ungnade, Herbert. *Guide to the New Mexico Mountains.* Albuquerque: University of New Mexico Press, 1965.

Visitor Guide: Sandia Mountains / Cibola National Forest. Albuquerque: Southwest Natural and Cultural Heritage Association, 1994.

Williams, Jerry L., ed. *New Mexico in Maps.* 2nd ed. Albuquerque: University of New Mexico Press, 1986.

Wilson, Alan, and Gene Dennison. *Navajo Place Names.* Guilford, CT: Jeffrey Norton, 1995.

Wolf, Tom. *Colorado's Sangre de Cristo Mountains.* Niwot, CO: University Press of Colorado, 1995.

Woodward, Lee. *Geology and Mineral Resources of Sierra Nacimiento.* Socorro: NM Bureau of Mines and Mineral Resources, 1987.

Young, John V. *The State Parks of New Mexico.* Albuquerque: University of New Mexico Press, 1984.

Zeller, Robert A., Jr. *Geology of Little Hatchet Mountains, Hidalgo and Grant Counties, NM.* Bulletin 96. Socorro: State Bureau of Mines and Mineral Resources, 1970.

Zim, Herbert S., and Paul R. Shaffer. *Rocks and Minerals: a Guide to Familiar Minerals, Gems, Ores, and Rocks.* New York: Golden Press, 1957.

Index

Note: Page numbers in *italic* text denote illustrations.